Second Revise

A **Dictionary** *of*
—— Modern ——
Star Names

A Short Guide to 254 Star Names and Their Derivations

Paul Kunitzsch and Tim Smart

Sky Publishing
A New Track Media Company
Cambridge, Massachusetts

© 2006 New Track Media LLC
Sky Publishing
49 Bay State Road
Cambridge, MA 02138-1200, USA
SkyTonight.com

Library of Congress Cataloging-in-Publication Data

Kunitzsch, Paul.
    [Short guide to modern star names and their derivations]
    A dictionary of modern star names : a short guide to 254 star names and their derivations / by Paul Kunitzsch and Tim Smart.
        p. cm.
    Originally published: Short guide to modern star names and their derivations. Wiesbaden : O. harrassowitz, 1986.
    Includes bibliographical references and index.
    ISBN-13: 978-1-931559-44-7 (pbk. : alk. paper)
    ISBN-10: 1-931559-44-9 (pbk. : alk. paper)
        1. Stars--Names--Dictionaries. 2. Constellations--Names--Dictionaries. I. Smart, Tim.
II. Title.

QB802.K818 2006
523.8--dc22

                                                                              2006049321

Originally published in 1986 as *Short Guide to Modern Star Names and Their Derivations,* by Harrassowitz Verlag, Wiesbaden, Germany.

Table of Contents

Introduction 1
Bibliographical Guide 13

CONSTELLATIONS

Andromeda	15	Grus	39
Aquarius	16	Hercules	39
Aquila	17	Hydra	40
Aries	18	Leo	40
Auriga	19	Lepus	43
Boötes	19	Libra	43
Cancer	21	Lyra	43
Canes Venatici	22	Ophiuchus	44
Canis Major	22	Orion	45
Canis Minor	24	Pavo	47
Capricornus	24	Pegasus	47
Carina	25	Perseus	48
Cassiopeia	26	Phoenix	49
Centaurus	27	Pisces	50
Cepheus	28	Piscis Austrinus	50
Cetus	29	Puppis	50
Columba	30	Sagitta	51
Corona Borealis	30	Sagittarius	51
Corvus	31	Scorpius	52
Crater	31	Serpens	54
Crux	32	Taurus	54
Cygnus	32	Triangulum	55
Delphinus	33	Triangulum Australe	55
Draco	34	Ursa Major	55
Equuleus	35	Ursa Minor	58
Eridanus	36	Vela	59
Gemini	37	Virgo	59

Appendix 62
Index 63

Paul Kunitzsch was professor of Arabic studies in the University of Munich, Germany, from 1975 until his retirement in 1995. His research concentrated on the transmission of science, especially astronomy, from ancient Greece to the Arabs and then to medieval Europe.

Tim Smart has been sharing his passion for the skies with students since 1978. He lives in California with his two children, Andromeda and Rigel.

This book was first published under the title *Short Guide to Modern Star Names and Their Derivations* by Harrassowitz Publishers in 1986.

Introduction

THE NAMES

Included in this paper are 254 star names, accompanied by their Greek letters or other modern designations. The names are arranged under their parent constellations, which are listed alphabetically in Latin as currently used in astronomy. To facilitate finding individual entries, the constellations and their page numbers are given in the Table of Contents, and an alphabetical Index of the names is provided at the end.

Certainly star names other than those listed here can be found on charts and in other books. This collection is not meant to be completely comprehensive. However, herein are included all the names commonly found in modern English sources, as well as a number of less common and even obscure names. While some of the present names can be found elsewhere with other spellings, here only the most common spellings found in modern astronomical sources have been selected. Also, several of the names or their variants can be found elsewhere applied to other stars; here again only the most common applications found in modern astronomical sources have been selected.

PRONUNCIATION

Each name is followed by its pronunciation. In most cases two pronunciations are shown. The one enclosed in parentheses (given first) is meant to be the approximate original pronunciation of the name, near to how it would have been spoken in its native tongue. The pronunciation enclosed in square brackets (given second) is meant to be the popular English pronunciation of the name, often differing significantly from the original. The symbols used in both sets of pronunciations are shown in Table 1. For reference purposes, Tables 2 and 3 show the proper pronunciations of Arabic and Latin vowels, used as guidelines in determining the approximate original pronunciations of the Arabic and Latin names.

With respect to the original pronunciations, for those names derived from Arabic or other Eastern languages, the sounds of the modern letters have been conformed as much as is reasonable to their original sounds and accents. However, through the process of transmission, many of the Arabic names have been misspelled, abbreviated, or their arrangement of syllables changed, making it difficult to determine a truly original pronunciation. In such cases erroneous consonants in the modern titles are mostly given their modern English sound. Erroneous vowels are treated similarly, but usually made long or short so as to "best conform" to a long or short sound in the different, original Arabic vowel.

The popular pronunciations are those that may be found in English reference works, or that may be commonly heard spoken. Still other popular pronunciations can probably be found in other books. As previously indicated, these pronunciations often differ significantly from the original Arabic (or even Latin and Greek) pronunciations, in many cases to the extent of being unrecognizable to a speaker of the native tongue.

There may be a question as to why one should offer new, "original" pronunciations, when established popular ones exist. The answer is that someone

Table 1. Pronunciation symbols used in this paper, including English and French vowels and diphthongs.

symbol	duration	examples	symbol	duration	examples
ə	murmur	abound, idea	ī	long	ice, time
ər	murmur	nature, baker	ŏ	short	offer, dog
ă	short	act, map	ö	long	beau, Bordeaux**
ä	long	arm, father	ō	long	only, bone
ā	long	aim, take	ô	long	all, jaw
â	long	air, care	ou	long	out, now
ě	short	end, best	ŭ	short	much, come
é	long	café, André*	u̇	short	put, good
ē	long	even, we	ū	long	clue, noon
ĭ	short	it, pin	yū	long	union, cube

* The French long e; pronounced close to English ā but without the "i" component at the end.

** The French long o; pronounced close to English ō but without the "u" component at the end.

interested in the original meanings of star names might also be interested in the original pronunciation of those names. Furthermore, an original pronunciation is, in a sense, more correct than a modern, fundamentally erroneous one. Perhaps our Arabic names "deserve" Arabic pronunciations? In any event the original pronunciations given here, as imperfect as many of them may be, are a rough attempt to show the pronunciations that were correct for these names in the places where they were originally used.

In many cases only one pronunciation is provided, usually because the original and popular pronunciations are approximately the same. It may also be that an obscure name has no "popular" pronunciation, or a very recent name

Table 2. Pronouncing Arabic vowels and diphthongs.*

Arabic transliteration	duration	approximate equivalent from Table 1	Arabic transliteration	duration	approximate equivalent from Table 1
a	short	ŭ	i	short	ĭ
ā	long	ä	ī	long	ē
ai	long	ā or ī	u	short	ủ
au	long	ou	ū	long	ū

* These equivalents are for basic sound values; in practice the pronunciation can be influenced by neighboring consonants.

Table 3. Pronouncing Latin vowels.

short Latin vowels	approximate equivalent from Table 1	long Latin vowels	approximate equivalent from Table 1
a	ŭ	a	ä
e	ĕ	e	é
i	ĭ	i	ē
o	ŏ	o	ö
u	ủ	u	ū
y	(no equivalent in English; like German ü or French u)	y	(no equivalent in English; like short y, but long)

has no "original" pronunciation. Where Arabic names have been severely distorted from their original form, no attempt has been made to give them original pronunciations, and only popular ones are provided.

DERIVATIONS

Following a name's pronunciation is its derivation – the original form or use of the word, its parent culture or language, its meaning (when known), and an abbreviated notice on its period of application as a Western star name. Throughout, the abbreviation i n d - A stands for "indigenous Arabic," and the abbreviation s c i - A stands for "scientific Arabic" (Greek-based, Arabic-Islamic). Classical Greek names are given in Greek letters. Arabic names are shown transliterated according to the current Anglophone orientalist system, rather than in original Arabic script. The Arabic alphabet thus becomes:

> *b t th j ḥ kh d dh r z s sh ṣ ḍ ṭ ẓ*
‹ *gh f q k l m n h w y a ā i ī u ū au ai* .

While most of the names have a long, detailed history of applications and modifications, such complete treatments are not given here. Generally only "results" – a beginning and an end of a name's history – are shown. However a few dates, names, and expanded derivations are included to give readers some feeling of the development of these names through time, at the hands of real people like ourselves.

HISTORY

Today's star names derive from a variety of past cultures and languages, mostly from the Middle East and the Mediterranean region (refer to Figure 1). Since medieval times they have been mixed together and passed

Figure 1. Cultural and language origins of today's star names, shown divided into Western and Eastern names. Bold arrows show mainstreams of transmission. Dotted arrows show unclear routes of transmission. Cultures which primarily had their own stellar nomenclature (to which may have been added foreign elements) are enclosed in boxes. Cultures which had a stellar nomenclature entirely made up of foreign elements are enclosed in square brackets. On the left margin appear the four chronological periods of formation and application of the names (not to scale). For a more thorough explanation of historical developments, refer to the text.

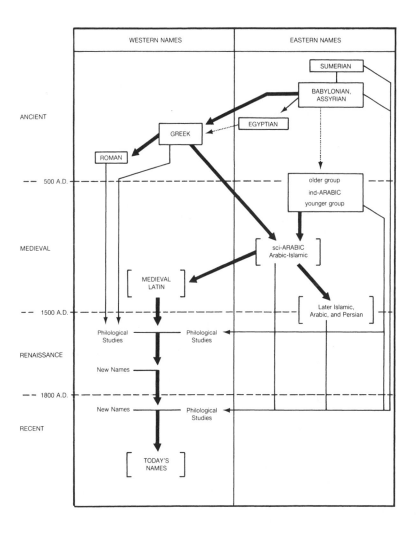

down to us in Latin characters. Their times of original formation or application can be grouped roughly into four historical periods, conveniently termed in this paper as follows:

1. *Ancient:* prior to 500 A.D.
2. *Medieval:* meant to encompass both Western and Eastern cultures coincident with Europe's medieval period, from approximately 500 A.D. to 1500 A.D.
3. *Renaissance:* meant to encompass the Renaissance period and the subsequent centuries, from approximately 1500 A.D. to 1800 A.D.
4. *Recent:* since 1800 A.D.

Ancient Times. Preeminent among the sources of ancient star names are the Greeks. Their astronomical nomenclature was established by 800 B.C. for the most prominent stellar objects, and by 400 B.C. for the majority of the 48 classical constellations. A final summing up of their names and astronomical knowledge was represented in Claudius Ptolemy's *The Great System of Astronomy* (today called by its shortened Arabic title, the *Almagest*), written ca. 150 A.D. in Alexandria, Egypt. In his book, Ptolemy catalogued 1,025 fixed stars, identifying each by its location in one of the 48 constellations, together with its ecliptic longitude and latitude, and its magnitude. In addition, several prominent stars bore proper Greek names (for example "Arcturus"). It is likely that many of the Greek constellations, and the names of some individual stars, derived from previous Babylonian or even older Sumerian originals. Babylonian astronomy and astrology influenced the Greeks down to the Hellenistic period, to as late as 100 to 200 A.D.; some Egyptian influence is also indicated. In turn, the Greeks passed on a portion of their star lore and nomenclature to the Romans.

Another source of ancient names are the inhabitants of the Arabic peninsula, mostly "desert nomads," or Bedouins. From these people about fifteen of their oldest names are known for the most prominent objects in the sky (for example "Aldebaran"). Here again Babylonian or Sumerian influences are sometimes indicated. Also, offshoots from the Babylonian zodiac seem somehow to have made their way to the ind-Arabs during this early time, influencing a number of their later star names. Further, a system of twenty-eight lunar mansions was somehow received from India, upon which were transferred pre-existing ind-A star names.

Medieval Times. The main body, or younger group, of ind-A names were probably formed in the period 500–700 A.D. In these later names, single stars stood for single animals, and were named individually, or in pairs or groups. Other stars were named for abstract qualities (for example "the Solitary One," α Hya). Less commonly several stars were grouped to form one object, in the manner of Western constellations. All these ind-A figures and names were created by different individuals or bands, in scattered places in Arabia and at different times. They were probably not too frequently used, and certainly not throughout all Arabia at once. Mixing in with the younger names at this point were the older names passed on through oral tradition. It was in the period following the spread of Islam (lifetime of the prophet Muḥammad: ca. 570–632 A.D.), that there was a stimulated use of writing in Arabia. As a result, poets and other authors began using and inflating the existing star names, and created new names for their own imaginative use.

Still later after the spread of Islam, a new center of intellectual activities arose away from the Arabian peninsula, in the new, northern capital of the caliphate: Baghdad (founded in 762 A.D.). At this point the ind-A culture gradually became superseded by the growing Arabic-Islamic (sci-A) culture. The rising power of the Arabs in the Middle East and beyond brought them exposure to the neighboring civilizations that they subjugated. Hence the developing Arabic-Islamic culture partly retained its Middle Eastern traditions, but, in the development of the sciences, largely adopted Western (Hellenistic) traditions passed along from Alexandria and other centers of Greek and Christian learning.

In the 9th and 10th centuries A.D., lexicographers of the classical Arabic language (such as Ibn Qutaiba), and sci-A astronomers (above all al-Ṣūfī) looked back at the ind-A heritage and collected some 350 star names from the old sources. Unfortunately the meanings of many of those names were no longer understandable and remained uncertain (as they remain today, though some conjectured meanings from the past ten centuries have often been taken as fact). Names or words from the old poetry were also often interpreted as authentic star names, resulting in a mixture of real and imagined names, and true origins became clouded. Furthermore, the ind-A names for far-southern stars (visible only deep in the Arabian peninsula), could not be satisfactorily understood nor assigned by the more northern sci-A compilers. Thus the Western world would later receive, from the sci-Arabs, a portion of the old, often confused, ind-A names.

Today, the ind-A names may appear as a single body of titles applied congruously to the sky, but that is a false impression. It must be remembered that they represent a collection of archaic names from different individuals, different areas, and different centuries. In many cases the sci-Arabs recorded several ind-A names for just one star or asterism, each from a different source. Just a sampling of these came into Western use. And while some sci-A compilers and later Westerners were prone to imagine connections between the disparate ind-A names, the results of such connections (as easy or natural as they can still be to make) were bound not to be authentic. The original sources themselves were far from being connected.

Despite the retention of a portion of the ind-A nomenclature, the practical astronomy of the sci-Arabs was largely Greek-based. Ptolemy's *Almagest* served as a foundation, being translated into Arabic several times in the 8th and 9th centuries, with resulting slightly different versions followed by later revisions. In the translations, the few Greek proper star names were mostly replaced by respective ind-A names. The images of the Greek constellations were faithfully retained (with some oriental adaptations), though their names were variously translated, transliterated, or replaced by a respective ind-A name. Ptolemy's descriptions, in the *Almagest*, of a star's location within a constellation – such as "the bright star in the left foot" (β Ori), "the star on the end of the tail" (β Leo), etc. – were converted as needed into proper star names by the sci-Arabs (or by subsequent Westerners), and so compose the main body of today's "sci-A" star names.

Assimilated with the varied names from the Arabic editions of the *Almagest*, were names from several other Greek astronomical and astrological works translated during the same period, including Ptolemy's astrological handbook, the *Tetrabiblos*. All these names were being adopted and used in innumerable sci-A works going well beyond the *Almagest*-works on astronomy and astrology, and as engravings on globes and astrolabes with their accompanying texts. Also appearing in these books were technical terms, some of them later to be confused as star names by Western astronomers (for example Giausar, λ Dra). Eventually, Arabic-Islamic authors writing in other Middle Eastern languages (such as the Persian Naṣīr al-Dīn al-Ṭūsī, 13th century; and the Mongol prince Ulugh Bēg, 15th century) translated and supplemented the previous sci-A works. Thus in many forms and from many sources the sci-A names, plus a collection of ind-A names, plus technical terms and a few other Middle Eastern names, would be transmitted to the West.

It was in Spain that the expanding Arabic-Islamic empire encountered

Europe set in the Middle Ages. There primarily, in the 10th to 13th centuries, works on sci-A astronomy, astrology, and instruments, including the *Almagest* (as well as books on alchemy, mathematics, medicine, and all else), were translated into Medieval Latin or Old Spanish. But it was a disunified process. The Arabic sources were of course exceedingly diverse, both in content and orthography. Also, the work of transferral (again a combination of translation and transliteration) was done by authors in different places and at different times. Often the work was done with the help of an Arabic speaker who was interpreting a source according to his own peculiar dialect. Different versions of many of the books were produced, and even the use of Latin was varied in its grammar and orthography during this period. There were copying errors (all books being tediously hand-copied), mistaken transfers of names, and some erroneous translations or misinterpretations of names. With such disunity in the stellar nomenclature, there was bound to be confusion for whomsoever in the future attempted to sort the subject out.

Renaissance Times. In the Latin borrowings of the 10th to 13th centuries, Arabic star names had been newly acquired or disseminated in the Western world rather continuously. The names had always come together with their texts, and were therefore integral parts of the astronomical or astrological subject matter. However, in the Renaissance period and the centuries that followed, the era of influx and translation was centuries past. The formerly novel star names had become part of a long-established stellar nomenclature, which was generally considered apart from its original, subject-bound contexts.

At this point two new developments in the history of star names occurred, each accompanied by its own sources of error. One of these developments was the appearance of philological studies, which examined and attempted to understand and explain the origins of the stellar nomenclature inherited from the past. The subject of these studies was of course the diverse Arabic and Latin names from the medieval period, but also ancient Greek and Roman names. The latter had survived the Dark Ages in the few, rediscovered classical books on astronomy and astrology remaining in libraries and other collections. These early philological studies, however, often suffered from a scarcity of original Arabic source material, and from insufficient knowledge of the necessary oriental languages on the part of the authors. Consequently, there was a frequent misunderstanding of those sources that were available, and plenty of erroneous speculation.

Another new development in Renaissance times came at the hands of astronomers. They deliberately searched the philological studies for new names to apply to their charts and globes. One example was Johannes Bayer, who, in his *Uranometria* of 1603, plundered the studies of Joseph Scaliger and Hugo Grotius published in 1600, in addition to selecting from the available source texts from the medieval translations and their subsequent reworkings, etc. However, the astronomers were evidently not sufficiently trained or interested in a scrupulous evaluation of the philological material. They therefore applied names, or mere words, from these sources according to their own will or understanding, adding much to the wrong uses of star names in modern astronomy.

In addition to borrowing from philological studies, the astronomers borrowed from other literary sources, or created their own new names. For example, names were invented to honor sovereigns ("Cor Caroli," for α CVn), or to describe stars ("Mira," for o Cet). Other names were invented or transferred to stars in the new constellations of the southern hemisphere, following the great exploratory voyages south of the equator.

Recent Times. The post-Renaissance period saw further philological studies, and continued borrowings by astronomers. One author who made an important contribution to the historical explanation of star names based on Arabic sources, was the German astronomer and historian Ludewig Ideler (1809). But while Ideler used some Arabic material, this was of a relatively secondary character, and not in a good state of transmission. Lacking a knowledge of the total breadth of past tradition and transmission, the results of Ideler's philological study were only partly correct, and subject to wrong conclusions or derivations in many cases. Nevertheless, his book remained the basic source for star name studies, especially in Arabic names, for the following 150 years.

The generally accepted star name study in the English language was Richard Hinckley Allen's *Star Names and Their Meanings* (1899; reprinted in 1963 with the slightly modified title *Star Names – Their Lore and Meaning*). In its Arabic material, this book was largely based on Ideler's. However, Allen added many new errors and faults of his own, both in matters of language and in interpretations of words and names. Equally, more recent English-language studies of star names were not dependable. For example the work of George A. Davis, Jr., "Pronunciations, Derivations, and Meanings of a Selected List of Star Names" (which appeared in *Popular Astronomy*, January, 1944; reprinted in 1963), contained a number of derivations that were not in accord with in-depth research into the original sources.

All these philological and historical dissertations were again productive in the formation of new Western star names. Astronomers continued to sort out single names and words from such studies and apply them as proper names to stars that had remained untitled so far. For example, early in the 19th century the Italian astronomer Giuseppe Piazzi (in his Palermo Catalogue of 1814) plundered the former philological study of Thomas Hyde (1665), introducing nearly a hundred new "Arabic" and other star names into modern astronomical use. Also in recent times, material from still other cultures came to be studied in greater detail. Thus several ancient Sumerian, Babylonian, Chinese, and other names were discovered, "translated," and tentatively reapplied to the sky (sometimes incorrectly). In addition, a scattering of other new names were introduced. (The most modern new names from the 20th century, such as Mimosa, Atria, etc., have not been researched as deeply here as the older names; this is mainly because it has not been possible to find out where, why, and by whom they were first introduced, due to the vast volume of modern literature.)

Today our star names can still be found variously spelled, variously derived, and various meanings for them given. While this confusion is not everywhere untangled, the reasons for its happening are very apparent historically. Without a complete command of the many original languages involved, without access to as many as possible of the centuries-old original works as well as the later studies, it is impossible for authors of this intricate subject to obtain perfectly accurate results. It was not until 1959 that a new research into the "Arabic" star names was published (by Paul Kunitzsch, in German – see the Bibliographical Guide), which put the Arabic matter on a firm basis using all the original sources involved, and tracing all the individual names step by step backwards to their ultimate source. The present paper is a modest attempt to bring forward, for the first time in the English language, many of the accurate results determined by such latest research.

A statistical analysis of the 254 names here presented reveals that (counting five double entries only once) 175 names (= 70%) are Arabic and 47 (= 19%) are Greek or Latin. Further, three names are Persian, two Hebrew, one Turkish, two English and three were recovered from old Mesopotamian material. 15 names were artificially fabricated in Renaissance and recent times. For one name, Kochab, it cannot yet safely be established whether it was originally Hebrew or Arabic.

SOURCES

This paper was primarily written by Tim Smart (eduactor in California, USA). All derivations, and a strict editing of the introductory matter, were supplied by Dr. Paul Kunitzsch (Institute for Semitic Studies, University of Munich, West Germany). Dr. Kunitzsch's historical understandings and definitive results, in turn, were obtained entirely from direct consultation and critical analysis of original source material – ancient, medieval, Renaissance, and younger, written in Arabic, Persian, Latin, Greek, and in the languages of modern research.

Dr. Kunitzsch also provided or helped with the original pronunciations of names. Sources for the popular pronunciations were the aforementioned paper of George A. Davis, Jr., lists of names in astronomical publications, and greater dictionaries of the English language.

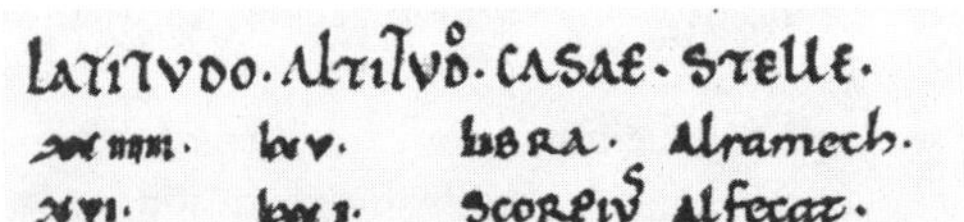

Fig. 2: Part of the oldest Latin star table with Arabic star names (late tenth century) in a manuscript copy of the eleventh century. Some of the names here introduced are still used today.

Bibliographical Guide

For readers who might wish to go deeper into details, the following books can be named as being authoritative for the celestial nomenclature of different cultures:

Sumerian and Babylonian (Akkadic): *Šumerisches Lexikon*, ed. P. Anton Deimel S.I., part iv, vol. 2: *Planetarium Babylonicum*, by P. Gössmann O.E.S.A., Rome 1950; H. Hunger – D. Pingree, *MUL.APIN, An Astronomical Compendium in Cuneiform* (Archiv für Orientforschung, Beiheft 24, 1989).

Indo-European Peoples: Anton Scherer, *Gestirnnamen bei den indogermanischen Völkern*, Heidelberg 1953 (including Greek and classical Latin).

Classical Roman: André le Boeuffle, *Les noms latins d'astres et de constellations*, Paris 1977.

Indigenous Arabic: Paul Kunitzsch, *Untersuchungen zur Sternnomenklatur der Araber*, Wiesbaden 1961. New materials and additions, by the same author: *Über eine anwāʾ-Tradition mit bisher unbekannten Sternnamen*, München 1983 (Bayerische Akademie der Wissenschaften, Sitzungsberichte).

Scientific Arabic (*Almagest*): Paul Kunitzsch, *Der Almagest. Die Syntaxis Mathematica des Claudius Ptolemäus in arabisch-lateinischer Überlieferung*, Wiesbaden 1974.

Arabic in Western Use: Paul Kunitzsch, *Arabische Sternnamen in Europa*, Wiesbaden 1959. Abundant material from medieval Western star tables is presented by Paul Kunitzsch, *Typen von Sternverzeichnissen in astronomischen Handschriften des zehnten bis vierzehnten Jahrhunderts*, Wiesbaden 1966.

Text of Ptolemy's star catalogue in the *Almagest*: Greek text edited by J. L. Heiberg, in vol. i, part ii, of Ptolemaeus, *Opera*, Leipzig 1903. An updated recent English translation of the *Almagest* is: *Ptolemy's Almagest*, Translated and Annotated by G. J. Toomer, London 1984. The star catalog of the *Almagest* has been edited in the two surviving Arabic translations and in Gerard of Cremona's medieval Latin translation from the Arabic, by Paul Kunitzsch: Claudius Ptolemäus, *Der Sternkatalog des Almagest. Die arabisch-mittelalterliche Tradition*, 3 vols., Wiesbaden 1986–1991.

ANDROMEDA (And)

α Alpheratz (ŭl fē' rŭts) [ăl fē' răts]

The specific origin of this name is unclear. It may represent a transfer from β Peg, where *alferaz* and other variants were applied in medieval times as abbreviations of β Peg's sci-A name *mankib al-faras,* "the Horse's Shoulder." Or, it may be from an abbreviation of the name *alpheraz id est equus,* "*alpheraz,* that is, the Horse," which was applied directly to α And also in medieval times. In any case, Medieval Latin authors confused these two names, their spellings, and their identifications, and the modern application of "Alpheratz" to α And comes down to us from late medieval times.

or Sirrah (sĭr' rŭ)

Applied in recent times from an abbreviation of its sci-A name *surrat al-faras,* "the Horse's Navel."

β Mirach [mī' răk]

Ultimately from the Arabic word *al-miʾzar,* "the girdle, or loin cloth," used in the Arabic *Almagest* in describing this star. The correct transliteration of the word in the Medieval Latin *Almagest* was *mizar,* which was occasionally misspelled as *mirac, mirat,* etc. These corruptions, taken as proper names for β And, were correctly attributed to *al-miʾzar* by Renaissance scholars. Subsequently one of the misspellings, "Mirach," gained more popularity over the correct "Mizar," to become the preferred modern name.

γ Almach (ŭl mäk') [ăl' măk]

The formation of this name begins with the ind-A name for this star: *ʿanāq al-arḍ,* "the Caracal" (a black-eared feline predator found in the Middle East). The ind-Arabs also gave the name in short form as *al-ʿanāq,* which became transliterated into Medieval Latin as *alamac.* In Renaissance times, the derivation of *alamac* was erroneously attributed to the assumed Arabic word *al-māq* (properly *al-mūq*), for "the boot, or buskin," rather than to *al-ʿanāq.* Subsequently the erroneous

word, as "Almaak" and later "Almach," was applied as a star name to
γ And, mostly in astronomical works in English. Other, non-English
works use the spelling "Alamak" derived directly from the Medieval
Latin transliteration.

ξ A d h i l (ŭ dāl′) [ə dĭl′]
From the Arabic word *al-dhail*, "the train of a robe or dress," used in
the Arabic *Almagest* in describing A and χ And, and transliterated in
the Medieval Latin *Almagest* as *adhil*. This word was wrongly applied
as a star name to ξ And in recent times.

AQUARIUS (Aqr)

α S a d a l m e l i k (säd′ ŭl mĕ′ lĭk)
Applied in recent times from the ind-A name *saᶜd al-malik*, for α and
o Aqr. A possible meaning for the name is "the Lucky (Stars) of the
King," but the exact historical connections are unknown.
Of the ten sets of stars in the region of today's Aquarius, Capricornus,
and Pegasus, whose ind-A names begin with the word *saᶜd*, none of
their meanings are really known. The knowledge was lost by the Arabs
themselves centuries ago. As a common noun in Arabic, *saᶜd* means
"luck." Furthermore, it has been suggested that all these stars may
have been associated with a pagan Arabic deity called Saᶜd.

β S a d a l s u u d (säd′ ŭl sù ūd′) [sŭd′ ăl sū′ ùd]
Applied in recent times from the ind-A lunar mansion name *saᶜd al-
suᶜūd*, for β and ξ Aqr, and 46 Cap. A possible meaning for the name
is "the Luckiest of the Lucky (Stars)," but the exact historical connections are unknown (see α Aqr).

γ S a d a c h b i a (säd ŭk′ bĭ ŭ) [sŭd ăk′ bĭ ə]
Applied in recent times from the ind-A lunar mansion name *saᶜd al-
akhbiya*, for γ, π, ζ, and η Aqr (today's "Y of Aquarius"). A possible
meaning for the name is "the Lucky (Stars) of the Tents," but the exact
historical connections are unknown (see α Aqr).

δ S k a t [skāt]

Applied with various spellings since medieval times, from the Arabic word *al-sāq*, "the shin," used in the Arabic *Almagest* in describing this star.

ε A l b a l i (ŭl bä′ lǐ) [ăl bä′ lē]

From the Arabic word *bāli*ʿ, "swallower," used in a sci-A discussion pertaining to the ind-A lunar mansion name *saʿd bula*ʿ (given for ε, μ, and ν Aqr; of unknown meaning [see α Aqr]). In recent times the Arabic article *al-* was added to *bāli*ʿ and the word was applied as a star name to ε Aqr.

θ A n c h a (ŭng′ kŭ)

A Latin word meaning "hip," used in the Medieval Latin *Almagest* in describing σ and ι/38 Aqr in the right and left hips, respectively (following a sci-A error, for Ptolemy had these stars in "the buttocks"). The word was applied as a star name to θ Aqr (correctly in Ptolemy's "right socket of the hip") in recent times.

κ S i t u l a (sĭ′ tù lŭ)

A Latin word meaning "pot, or bucket," used in Renaissance philological studies as the translation of the sci-A constellation name *al-dalw*, for Aquarius. Subsequently the word was applied as a star name to κ Aqr.

The ind-Arabs located *al-dalw* ("the Well Bucket") in today's Square of Pegasus. In that location it corresponded to what in other cultures was Aquarius among the zodiacal signs. The sci-Arabs subsequently used the indigenous name *al-dalw* for the Greek Water Pourer (Aquarius). An alternative sci-A name for Aquarius was the translation *sākib al-mā*ʾ, "the Water Pourer."

AQUILA (Aql)

α A l t a i r (ŭl tä′ ĭr) [ăl târ′]

Applied with various spellings since medieval times, from an abbreviation of its ind-A name *al-nasr al-ṭā*ʾir, "the Flying Eagle (or Vulture)," alternatively used as an asterism name for α, β, and γ Aql. The name

has probable origins among the Babylonians and Sumerians, for whom
α Aql was "the Eagle Star."

The outstretched or "flying" eagle configuration of α, β, and γ Aql
was seen by the ind-Arabs in contrast to the nearby close-winged,
"swooping" eagle configuration of α, ε, and ζ Lyr. It appears that the
ind-Arabs received the original idea of an eagle for α Aql, and later
divided that idea between α Aql and α Lyr.

β Alshain (ŭl shä ēn') [ăl shān']
γ Tarazed (tä rä zĕd') [tä' rə zĕd]

Both applied in recent times (with a misreading in the second word)
from abbreviating the Persian asterism name *shāhīn-i tarāzū*, "the
Scale Beam," for α, β, and γ Aql. The Persian name, in turn, was a
medieval translation of these stars' ind-A name *al-mīzān*, "the
Balance" (said to be a popular name for the ind-Arabs' *al-nasr al-ṭāʾir*
listed under α Aql).

ARIES (Ari)

α Hamal (hŭ' mŭl) [hă' məl]

Applied in recent times from the sci-A constellation name *al-ḥamal*,
"the Lamb," for Aries. *al-ḥamal* seems to belong to those zodiacal
constellation names already known in ind-A times. (Occasionally both
the ind-A and sci-A figures were called *al-kabsh*, "the Ram.")

β Sheratan (shĕ rŭ tän') [shĕ' rə tăn]

Applied in recent times from the ind-A lunar mansion name *al-sharaṭān*, for β and γ Ari. The name means "two" of something, but the
complete meaning is uncertain. Some sci-A authors suggested that it
meant "the Two Signs," implying these stars as some kind of celestial
indicator (being the first of the twenty-eight ind-A lunar mansions);
others assumed it meant "the Two Horns," referring to the ind-A
figure of *al-ḥamal* here (see α Ari and β Tau).

γ Mesarthim (mə sär tēm') [mĕ' zär tĭm]

The formation of this name begins with the ind-A lunar mansion name
al-sharaṭān, for β and γ Ari (see β Ari). From the numerous medieval

lists of the ind-A lunar mansions (all in Latin transliteration), the
Renaissance scholar J. Bayer grasped the form *Sartai*, and used it (in a
note appended to γ Ari) for the three brighter stars α, β, and γ Ari.
Bayer also erroneously explained *Sartai* as from the Hebrew word
mᵉshārᵉthīm, "servants" (well-known as a technical term in Hebrew
grammar), rather than from *al-sharaṭān*. Subsequently Bayer's erroneous word, written as "Mesarthim," was applied as a star name to γ Ari.

δ B o t e i n (bŏ tān′)
> Applied in recent times from the ind-A lunar mansion name *al-buṭain*,
> "the Little Belly," for δ, ε, and ϱ Ari.

AURIGA (Aur)

α C a p e l l a (kŭ pĕl′ lŭ)
> Its ancient Roman name meaning "the She-goat" (but more commonly
> given, in antiquity, as *Capra*), after the star's Greek name Αἴξ, "the
> Goat." Reapplied in recent times.
> The Roman names Capella, Aselli (for γ/δ Cnc), and others in the
> diminutive form (also in Greek), are likely meant to indicate an atypi-
> cal use of words, as an animal's name given to a star instead of to the
> animal (rather than meaning "small" animals or personages).

β M e n k a l i n a n (mĕn′ kŭl ĭ nän′) [mĕn kă′ lĭ năn]
> Applied in recent times from its sci-A name *mankib dhī 'l-ʿinān*, "the
> Shoulder of the Reinholder."

BOÖTES (Boo)

α A r c t u r u s (ŭrk tū′ rùs)
> From its ancient Greek name Ἀρκτοῦρος, "the Bear Watcher, or
> Guardian," referring to the nearby bear constellation (Ursa Maior).
> Reapplied in Renaissance times. An alternative meaning for the name
> is "Guardian of the North," where the Greek word ἄρκτος, "bear,"
> also came to mean "north," by its association with the northerly celes-
> tial bear.

β Nekkar (něk kär′) [něk′ kär]

Applied in recent times from a misreading of the sci-A constellation name *al-baqqār*, "the Ox-driver," for Boötes.

γ Seginus [sě jī′ nəs]

The formation of this name begins with the Greek constellation name Βοώτης (Boötes), which was transliterated and then corrupted in the manuscripts of the Arabic *Almagest*. One of these Arabic corruptions, in turn, was transliterated into Latin as *theguius*, which became further corrupted into *cheguius, ceginus*, etc. One form, *Ceginus*, was applied as a star name to γ Boo by late medieval times, and "Seginus" is its recent spelling.

ε Izar (ĭ zär′) [ī′ zär]

Applied in recent times from the Arabic word *izār*, "girdle, or loin cloth," being a later version of the original term *al-miʾzar* used in the Arabic *Almagest* in describing this star.

or Pulcherrima (pŭl kĕr′ rĭ mŭ)

Its recent Latin name meaning "the Most Beautiful," referring to this star's colorful duplicity in the telescope.

η Muphrid (mŭf′ rĭd)

The formation of this name begins with the ancient ind-A name for α Boo: *al-simāk al-rāmiḥ*, "the Lance-bearing *simāk*" (the meaning of *simāk* is uncertain). It seems that later ind-A poets expanded upon this name by making mention of some separate "lance" (*al-rumḥ*) that accompanied the star α Boo. Subsequent sci-Arabs attempted to identify this fictitious lance with actual stars, sometimes saying it was η Boo with nearby stars, sometimes saying it was "η Boo alone" (where "alone," in Arabic, is written *mufradan*). From these discussions came, with a slight copying mistake in the Arabic, the erroneous ind-A name *mufrad al-rāmiḥ*, roughly translating as "the Isolated Single One of the Lance-bearer." With a wrong vocalization and with abbreviation, this erroneous name, as "Muphrid," was applied to η Boo in recent times.

μ Alkalurops (ŭl kŭ lū′ rŏps)

The formation of this name begins with one of the words used by Ptolemy to describe this star in the *Almagest*: κολλόροβος, "club," rarely used as "shepherd's staff." This word was transliterated into Arabic as *qulūrūbus*, and hence into Latin as *calurus*. Then in Renaissance times, the derivation of *calurus* was mistakenly attributed to another Greek word: καλαῦϱοψ, "shepherd's staff." This erroneous word, in its turn, was transliterated into Latin, then Arabicized with the article *al-*, then its spelling corrupted, to become "Alkalurops," which was thereafter applied as a star name to μ Boo.

h M e r g a (mĕr′ gŭ)

A Latin word meaning "reaping hook," used in a Renaissance discussion of the constellation Boötes. According to this discussion, some classical sources mentioned a reaping hook held in Boötes' hand opposite the staff. Subsequently the word was applied as a star name to h Boo.

CANCER (Cnc)

α A c u b e n s [ă′ kyū bĕnz]

Applied with various spellings since medieval times, from the Arabic word *al-zubānā*, "the claw," used in Ptolemy's *Tetrabiblos* in describing the stars α and ι Cnc. Otherwise, the same Arabic word was used for the "claws" of Scorpius, cf. α Lib and β Sco (Graffias).

or S e r t a n (sĕr tän′) [sər′ tăn]

Applied in recent times from the sci-A constellation name *al-saraṭān*, "the Crab," for Cancer.

γ A s e l l u s B o r e a l i s (ŭ sĕl′ lŭs – bŏ′ rĕ ă′ lĭs) [– bŏ′ rē ă′ lĭs]
δ A s e l l u s A u s t r a l i s (ŭ sĕl′ lŭs – ous trä′ lĭs) [– ôs trä′ lĭs]

From their joint Roman name the *Aselli* (and also *Asini*), "the Asses, or Donkeys," being a translation of these stars' Greek name οἱ ῎Ονοι. Reapplied in Renaissance times, along with the Latin distinctions of "northern" and "southern." (See also α Aur regarding diminutives.)

ζ T e g m i n e (tég′ mĭ nĕ)

A Latin word in the ablative case meaning "covering, or shell," used in a Renaissance discussion of the constellation Cancer. The word was subsequently applied as a star name to ζ Cnc (but as a name it should be "Tegmen," in the nominative case).

CANES VENATICI (CVn)

α Cor Caroli (kŏr kä' rŏ lē) [kŏr kă' rŏ lī]
This Latin name is independent of the constellation CVn (introduced by
Hevelius, 1690). It first appeared in fuller form as Cor Caroli Regis
Martyris, "Heart of Charles, the Martyr King", on English star maps
since 1673 in honour of King Charles I of England, who was beheaded in
1649.

β Chara (kŭ' rä) [kā' rə]
Applied in Renaissance times from the Greek word χαρά, "joy," that
was used by Hevelius in 1690 to name the southern dog (marked by α
and β CVn) in his new constellation Canes Venatici. ("Asterion,"
from the mythological Greek name Ἀστέριον, "Little Star," was the
name Hevelius gave to the northern dog [marked by 18, 19, 20, and 23
CVn].)

CANIS MAJOR (CMa)

α Sirius (sē' rĭ ŭs)
From its ancient Greek name Σείριος, "the Scorching One, or Bril-
liant One," appropriate for this brightest of the fixed stars. Reapplied
in Renaissance times.

β Mirzam (mĭr' zŭm)
Applied in recent times from its ind-A name al-mirzam, of unknown
meaning. The ind-Arabs also gave the name to β CMi, and sometimes
to γ Ori. Each al-mirzam star preceded the rising of a brighter star
(our Sirius, Procyon, and Betelgeuse, respectively), which probably
connects to the unknown meaning.

γ Muliphein (mù lĭ fān')
The formation of this name begins with the ind-A names ḥaḍāri and
al-wazn (see δ CMa). In Arabic discussions of these names it came to
be said that they were muḫlifān, meaning "two [things] causing dis-
pute [in this case as to these stars' identification] and the swearing of an
oath." From this usage, al-muḫlifān (then provided with the article
al-) was wrongly taken as a name for ḥaḍāri and al-wazn together.
Transliterated and further mutilated as "Muliphein," the name was

arbitrarily applied to the single star γ CMa in recent times. (For a subsequent transfer with a modernized spelling, see γ Cen.)

δ W e z e n (wĕ′ zən)

From some ind-A name *al-wazn*, for one of a pair of stars, the other one being *ḥaḍāri*. Sci-A authors ventured to identify these stars as α/β Cen or α/β Col, but exactly what two stars were originally intended, and the significance of their names, is unknown. As a common noun in Arabic, *al-wazn* means "the weight." "Wezen" was arbitrarily applied to δ CMa in recent times, and subsequently to β Col as well. (See also ζ CMa, γ CMa, and β Cen.)

ε A d h a r a (ŭ dä′ rŭ) [ə dā′ rŭ]

Applied in recent times from the ind-A asterism name *al-ʿadhārā*, "the Virgins," for ε, δ, η, and o² CMa (and perhaps a fifth unidentified star). The significance of the name is unknown. (See also η CMa.)

ζ F u r u d (fù rūd′) [fū′ rūd]

Ultimately from the Arabic word *al-furūd*, "the solitary ones," used in an ind-A poet's allusion to anonymous "solitary stars" around some star *ḥaḍāri*. The word was not intended as a name for specific stars as some later sci-A authors supposed. The attempted identity of the *"al-furūd"* stars, like their companion *ḥaḍāri* (see β Cen), was variously attributed to stars in today's Centaurus and Columba. The latter stars were included under Canis Maior in the *Almagest*, leading, in recent times, to the confused assignment of the erroneous "Furud" to ζ CMa.

η A l u d r a (ŭl ùd′ rŭ)

Applied in recent times from the alternative ind-A asterism names *al-ʿudhra*, "the Virginity," and *ʿudhrat al-jauzāʾ*, "the Virginity [?] of *al-jauzāʾ*," for ε, δ, η, and o² CMa (a third alternative for these stars was *al-ʿadhārā* – see ε CMa). The name *ʿudhrat al-jauzāʾ*, by tying in with the nearby ind-A asterism *al-jauzāʾ* (today's Orion), may have been an attempt to apply significance to the enigmatic and perhaps older name *al-ʿudhra*.

CANIS MINOR (CMi)

α Procyon (prŏ′ kĭ ŏn) [prō′ sĭ ŏn]

From its ancient Greek name Προκύων, "the One Preceding the Dog," referring to its rising shortly before the "Dog Star," Sirius. Reapplied in Renaissance times.

β Gomeisa (gŏ mā sä′) [gŏ mī′ zə]

From the ind-A surname al-ghumaiṣāʾ, "the Little Bleary-eyed One ('with a filthy fluid in the corner of the eye')," for α CMi. Wrongly transferred to β CMi in recent times. The original significance of the surname is unknown.

Perhaps as an attempt to explain the significance of al-ghumaiṣāʾ, as well as that of the surname al-ʿabūr, "the One Having Crossed Over [a river, etc.]," for α CMa, an Arabic fable developed associating these and other equally enigmatic star names. According to one version of the fable, al-ghumaiṣāʾ and al-ʿabūr were sisters, and their brother was suhail (α Car). Suhail, in turn, was the suitor of al-jauzāʾ (the feminine ind-A figure in place of Orion). In coitus, suhail broke the spine of al-jauzāʾ, thus killing her, after which suhail fled south. He was followed by his sister al-ʿabūr who "crossed over" the Milky Way (where the two stars now lie in the southern sky). Meanwhile suhail's second sister, al-ghumaiṣāʾ, was left alone north of the Milky Way, weeping, until her "eyes became bleary."

(The ind-A name for α CMa was al-shiʿrā, of unknown meaning. Apparently it was developed into a dualis form, al-shiʿrayān, to accomodate both α CMa and α CMi, with the aforementioned "surnames" distinguishing the two.)

CAPRICORNUS (Cap)

α Algedi (ŭl jĕ′ dē) [ăl jē′ dē]

Reapplied in recent times (from its older medieval spelling), from the sci-A constellation name al-jady, "the Kid," for Capricornus.

or Giedi (jĕ′ dē)

Applied in recent times, in a recent spelling, from the sci-A constellation name al-jady, "the Kid," for Capricornus.

β D a b i h (dä′ bĭ) [dā′ bē]

Applied in recent times from the ind-A lunar mansion name *sa'd al-dhābiḥ*, for α and β Cap. Possible meanings for the name are "the Lucky (Stars) of the Slaughterer," or, "Sa'd, the Slaughtering One" (this last using Sa'd as an unspecified proper name). However the exact historical connections are unknown (see α Aqr).

γ N a s h i r a (nä′ shĭ rŭ) [nā′ shĭ rə]

Applied in recent times from the ind-A name *sa'd nāshira*, for γ and δ Cap. Its meaning is unknown (see α Aqr). As originally applied, γ was "Nashira Prima" and δ was "Nashira Posterior."

δ D e n e b A l g e d i (dĕ′ nĕb – ŭl jĕ′ dē) [– ăl jē′ dē]

Applied since medieval times from its sci-A name *dhanab al-jady*, "the Kid's Tail," variously for γ or δ Cap, since both stars are located on the "tail" in the *Almagest*.

CARINA (Car)

α C a n o p u s (kŭ nö′ pŭs)

From its ancient Greek name Κάνωβος (Ptolemy's spelling), an untranslated proper name that was introduced rather late into Greek astronomy (perhaps in the 2nd century B.C.). There seem to be Egyptian influences in the name's development. Reapplied in Renaissance times.

or S u h e l (sŭ hāl′)

Applied in medieval times from its ind-A name *suhail*, of unknown meaning. (For a related borrowing with a modernized spelling, see λ Vel.)

β M i a p l a c i d u s [mī′ ə plă′ sĭ dəs]

Applied in recent times, and of unknown astronomical significance. The first element, Mia-, is unexplained. The second element is the Latin adjective *placidus*, meaning "calm, gentle" (here in the masculine form).

ε A v i o r [ă′ vĭ ôr]
>   Applied in recent times, and of unknown derivation. Perhaps it is the
>   reverse spelling of someone's name, "Roiva" (as in the case of α and
>   β Del, and γ Vel).

ι A s p i d i s k e (ŭs′ pĭ dĭs′ ké)
>   Applied in recent times from the Greek word ἀσπιδίσκη, "little
>   shield," used by Ptolemy in the *Almagest* in describing several stars in
>   his constellation Argo, that were fixed to the ship for protection and
>   decoration. Ptolemy's shields were in today's Puppis and Vela (hence
>   they did not include ι Car), but their identification was confused when
>   Argo was divided into the modern Carina, Puppis, Vela, and Pyxis.
>   (See also ϱ Pup.)

CASSIOPEIA (Cas)

α S h e d a r (shĕ′ dər)
>   Applied with various spellings since medieval times, from the Arabic
>   word *al-ṣadr*, "the breast," used in the Arabic *Almagest* in describing
>   this star.

β C a p h (kŭf) [kăf]
>   Applied in recent times from an abbreviation of its ind-A name *al-kaff
>   al-khaḍīb*, "the Stained Hand," alternatively and more correctly used
>   for all the brighter stars of today's Cassiopeia (probably α, β, γ, δ, and
>   ε). The ind-A figure here represented a hand with its finger-tips
>   stained reddish-brown in the traditional Eastern way using henna
>   leaves.
>   *al-kaff al-khaḍīb*, in turn, was part of the larger ind-A asterism *kaff al-
>   thurayyā al-yumnā al-mabsūṭa*, "the Outstretched Right Hand of the
>   Pleiades." The latter extended from today's Taurus through Perseus
>   into Cassiopeia. A second hand (the Pleiades were a "head" from
>   which two arms or hands radiated) was *al-kaff al-jadhmāʾ*, "the
>   Amputated Hand," in today's Cetus.

δ R u c h b a h (rŭk′ bŭ)
>   Applied in recent times from an abbreviation of its sci-A name *rukbat
>   dhāt al-kursīy*, "the Knee of the Lady of the Chair."

CENTAURUS (Cen)

α Rigil Kentaurus (rǐ′ jəl – kĕn tou′ rùs) [rī′ jǐl – kĕn tô′ rəs]

Applied in recent times from its sci-A name *rijl qanṭūris*, "the Centaur's Foot." Today the name is occasionally seen abbreviated as "Rigil Kent."

or Toliman (tŏ lē män′) [tō′ lǐ mǎn]

Applied in recent times from some ind-A name *al-ẓulmān*, "the Ostriches." Stars in today's Centaurus were variously identified with one or more ostriches by the ind-Arabs, but what star or stars were originally designated is unknown.

or Bungula [bùng′ gū lə]

Applied in recent times and probably coined from the Greek word *beta*, plus the Latin word *ungula*, "hoof" (which seems to be a substitution for "foot," of the Centaur, which describes α Cen in the *Almagest*). However, the initial letter "B," for the designation *beta*, is not correct for the star.

α C Proxima (prŏk′ sǐ mǔ)

Its recent Latin name meaning "the Nearest," as this component of the α Cen system is currently the nearest star to our solar system (it is visible only in telescopes).

β Hadar (hǔ där′) [hǎ′ dər]

From some ind-A name *ḥaḍāri* (an untranslated proper name), for one of a pair of stars, the other one being *al-wazn*. Sci-A authors ventured to identify these stars as α/β Cen or α/β Col, but exactly what two stars were originally intended, and the significance of their names, is unknown. "Hadar" was arbitrarily applied to β Cen in recent times. (See also γ, δ, and ζ CMa.)

or Agena [ə jē′ nə]

Applied in recent times and probably coined from the Greek word *alpha*, plus the Latin word *genu*, "knee" (Ptolemy described this star on the "knee" of the Centaur's left front leg). However, the initial letter "A," for the designation *alpha*, is not correct for the star.

γ Muhlifain (mù lǐ fān')

An erroneous ind-A name recently transferred, with this modernized
spelling, from a discussion of the name "Muliphein" for γ CMa (see
that star).

ζ Alnair (ŭl nä' ĭr) [ăl nâr']

Applied in recent times from an abbreviation of its sci-A name *nayyir
badan qanṭūris*, "the Bright One in the Body of the Centaur." Alnair
was taken from a wrong transliteration (Al Nā'ir) of the Arabic word
nayyir ("Bright"); cf. α Gru and α Phe.

θ Menkent (mĕn' kĕnt)

Applied in recent times and possibly coined from the Arabic word
mankib, "shoulder" (in some recent transliterations: *menkib*; Ptolemy
described this star on the right "shoulder" of the Centaur), plus the
Latinized "Kentaurus" for the sci-A constellation name *qanṭūris*.
Hence this name could be of similar construction to the abbreviation
for α Cen: "Rigil Kent."

CEPHEUS (Cep)

α Alderamin (ŭl' dĕ rä mēn') [ăl dĕ' rə mĭn]

Applied to α Cep since medieval times. The derivation understood
since Renaissance times was from *al-dhirāʿ (al-)yamīn,* "the Right
Forearm," an assumed sci-A descriptive term for α Cep. But this is
apparently erroneous. Firstly, in the *Almagest,* α Cep is described on
the right "shoulder" (Arabic *al-katif* or *al-mankib*) of Cepheus,
whereas no right "forearm" (*al-dhirāʿ*) is described for the figure.
Secondly, in Arabic, "right forearm" is spelled correctly as *al-dhirāʿ
al-yumnā,* which (in contrast to *al-yamīn*) lacks assonance in the last
syllable to Alderamin.

Therefore, it is more likely that Alderamin is derived from the various
medieval Western abbreviations of *muqaddam al-dhirāʿain,* "the Preceding One of the Two Cubits, or Forearms," which was a sci-A name
(based on ind-A) for α Gem. This would have been wrongly transferred to α Cep in medieval times.

β Alfirk (ŭl fĭrk′) [ăl′ fərk]

Applied in recent times from an abbreviation of the ind-A names *kaukabā al-firq* (for α and β Cep), and *kawākib al-firq* (for α, β, and η Cep). *kaukabā* and *kawākib* mean "two stars" and "stars" (more than two), respectively, but the original Arabic script for *al-firq* can be vocalized in various ways with various meanings. If the vocalization *al-firq* is accepted, it means "the Flock," and may refer to a flock of sheep here (see γ Cep).

γ Errai (ĕr rä′ ē) [ər rä′ ē]

Applied in recent times from its ind-A name *al-rāʿī*, "the Shepherd." This shepherd, with the sheep he attended (*al-aghnām*, marked by the various dim stars nearby), and the shepherd's dog (*kalb al-rāʿī*, marked by 28/29 [ρ] Cep), seem to form a complete group of ind-A figures. (Compare to a different group discussed under β Oph.)

μ "The Garnet Star"

Its recent English name, descriptive of this star's deep red color.

ξ Kurhah (kŭr′ hŭ)

Applied in recent times from one vocalization of its ind-A name: *al-qurḥa*, "the White Spot on the Forehead of a Horse." However, the original Arabic script for the name can be read and vocalized in other ways with various meanings.

CETUS (Cet)

α Menkar (mĕn′ kŭr) [mĕn′ kär]

From the Arabic word *al-minkhar*, "the nostrils," used in the Arabic *Almagest* in describing λ Cet. Wrongly applied as a star name to α Cet (properly on the "jaw") in a Medieval Latin star table.

β Diphda (dĭf′ dŭ)

Applied in recent times from an abbreviation of its ind-A name *al-ḍifdiʿ al-thānī*, "the Second Frog" (α PsA was the ind-Arabs' "First Frog," *al-ḍifdiʿ al-awwal*).

or D e n e b K a i t o s (dĕ' nĕb – kī' tŏs) [– kā' tŏs]
Applied in medieval times from an abbreviation of its sci-A name
dhanab qaiṭus (al-janūbī), "(the Southern [Branch] of) the Sea Monster's Tail."

ζ B a t e n K a i t o s (bŭ' tən – kī' tŏs) [bā' tən – kā' tŏs]
Applied in recent times from its sci-A name *baṭn qaiṭus*, "the Sea
Monster's Belly."

o M i r a (mē' rŭ) [mī' rə]
Its Latin name since Renaissance times meaning "the Amazing One,"
as taken from *Historiola Mirae Stellae* – the title of a work describing
the amazing variability of this star (written by Hevelius in 1662).

COLUMBA (Col)

α P h a c t (făkt) [făct]
Applied in recent times from the Arabic word *fākhita*, "ring dove,"
used in connection with the constellation Cygnus in a Renaissance
discussion of Arabic bird names.

β W a z n (wŭzn)
Applied in recent times from some ind-A name *al-wazn*, of unknown
significance (see δ CMa).

CORONA BOREALIS (CrB)

α A l p h e c c a (ŭl fĕk' kŭ)
Applied in medieval times from the ind-A asterism name *al-fakka*, for
Corona Borealis. The Arabic name, from the root *f-k-k*, "to separate,
break up, etc.," apparently refers to this asterism's shape: an incom-
plete circle.

or G e m m a (gĕm' mŭ) [jĕm' mə]
Its recent Latin name meaning "the Gem, or Jewel." The name may be
taken from one Renaissance author's discussion of the southern con-

stellation Corona Australis, which was described, in Latin, as "decorating mid-heaven [i.e., culminating] with its jewels [Latin *gemmis*, in the ablative plural], or stars, in early July." No other possible source for the name has been found.

β N u s a k a n (nù sŭ kän') [nū' sə kăn]
Applied in recent times from the collective ind-A name *al-nasaqān*, "the Two Lines [of stars]," for two asterisms in today's Hercules, Serpens, Ophiuchus, and Lyra.
al-nasaqān originally marked the boundaries of the ind-Arabs' *al-rauḍa*, "the Pasture" (see β Oph). These boundaries were "the Northern Line," *al-nasaq al-shaʾāmī* (including κ, γ, β, δ, λ, μ, o, ν, and ξ Her; β and γ Lyr; and β and γ Ser), and "the Southern Line," *al-nasaq al-yamānī* (including δ, λ, α, and ε Ser; and δ, ε, υ, η, ζ, and ξ Oph).

CORVUS (Crv)

α A l c h i b a (ŭl kĭ bä') [ăl kē' bə]
From the ind-A asterism name *al-khibāʾ*, "the Tent," for β, γ, δ, and ε Crv. Wrongly applied as a star name to α Crv in recent times.

γ G i e n a h (jĕ nä') [jē' nə]
Applied in recent times from an abbreviation of its sci-A name *janāḥ al-ghurāb*, "the Raven's Wing."

δ A l g o r a b (ŭl gŏ räb') [ăl gō' răb]
From an abbreviation of the sci-A name *janāḥ al-ghurāb*, "the Raven's Wing," for γ Crv. Transferred to δ Crv in Renaissance times (both γ and δ were on the Raven's wings in the *Almagest*).

CRATER (Crt)

α A l k e s (ŭl kés') [ăl' kĕz]
Applied with various spellings since medieval times, from the sci-A constellation name *al-kaʾs*, "the (Wine) Cup," for Crater.

CRUX (Cru)

α Acrux [ā′ krŭks]

Applied in recent times and obviously coined from its Greek letter designation *alpha*, plus the constellation name Crux (which in the genitive should be Crucis).

β Mimosa (mē mö′ sŭ)

Applied in recent times, and of unknown astronomical significance. It is from the Latin word *mimus*, "an actor," and is otherwise known as the genus name for certain tropical plants.

γ Gacrux [gŭ′ krŭks]

Applied in recent times and obviously coined from its Greek letter designation *gamma*, plus the constellation name Crux (which in the genitive should be Crucis).

Gacrux, and several other star names applied since the 19th century (those for ε Car, α/β Cen, α Cru, α TrA, γ Vel, and perhaps others), were probably invented by ocean navigators in need of proper names for the brighter southern stars.

CYGNUS (Cyg)

α Deneb (dĕ′ nĕb)

Applied with various spellings since medieval times, from an abbreviation of its sci-A name *dhanab al-dajāja*, "the Hen's Tail."

or Arided [ă′ rĭ dĕd]

Applied with various spellings since medieval times, from its ind-A name *al-ridf*, "the One Sitting Behind the Rider (on the same animal)," or simply "the Follower" (here perhaps with regard to the four stars δ, γ, ε, and ζ Cyg, called *al-fawāris*, "the Riders").

β Albireo (ŭl bē′ rĕ ö) [ăl bĭ′ rē ō]

The formation of this name begins with Ptolemy's name for the constellation Cygnus: Ὄρνις, "the Bird." The sci-Arabs transliterated this name as *ūrnis*. The Medieval Latin translator of the Arabic *Almagest*, in turn, did not recognize any Greek word behind *ūrnis* (or whatever Arabic corruption he may have read), so he merely transliterated it into a form which appeared in the manuscripts variously as *eurisim, eirisun, eirisim,* etc. In a note appended to this Medieval Latin constellation name for Cygnus, one Latin commentator ventured to derive it from the name of an aromatic herb he knew: *ireus.* This erroneous commentary read, in brief part: *"eirisim... ab ireo,"* or, "[the constellation name] *eirisim...* [coming] from [the word] *ireus."* (In the Latin text, *ireus* is given as *ireo* in the appropriate ablative case.) Apparently in one manuscript of the Latin *Almagest,* the final words of this commentary, *ab ireo,* were written on the next line below the constellation title, where the descriptions of the stars begin. Since the first star entered in the *Almagest* under the constellation Cygnus is β Cyg, the words *ab ireo* seem to have been confused as a name for this star. "Arabicized" with the insertion of the letter "l," *ab ireo,* written as "Albireo," was applied as a star name to β Cyg by Renaissance times.

γ S a d r (sŭdr) [sădr]
Applied in recent times from an abbreviation of its sci-A name *ṣadr al-dajāja,* "the Hen's Breast."

ε G i e n a h (jĕ nä′) [jē′ nə]
A name transferred from γ Crv in recent times, where the Arabic word *janāḥ* means "wing" (in the *Almagest,* both ε Cyg and γ Crv lie on the wings of their respective constellations; see γ Crv).

π¹ A z e l f a f a g e [ŭ zĕl′ fə fäj]
From the sci-A constellation name *al-sulaḥfāt,* "the Tortoise," for Lyra (corresponding to the Greek image of a tortoise shell for the Lyre's soundboard). Transliterated as *Azelfage,* a Medieval Latin translator erroneously attributed this constellation name to Cygnus. Then in Renaissance times, the corruption "Azelfafage" was applied as a star name to π¹ Cyg. (See also γ Lyr.)

DELPHINUS (Del)

α S u a l o c i n [swä′ lō sĭn]
β R o t a n e v [rō′ tə nĕv]
Their recent names from Piazzi's Palermo Catalogue (1814). Read backwards they are "Nicolaus Venator," the Latin form of the Italian

name Niccolo Cacciatore. Cacciatore was Piazzi's assistant and successor, of great help to Piazzi in his later years when the astronomer nearly lost his sight due to an eye illness. (In Piazzi's spelling, α was "Svalocin.")

DRACO (Dra)

α Thuban (thù bän') [thū' băn]
:   The formation of this name begins with the sci-A name for γ Dra: *ra^)s al-tinnīn*, "the Serpent's Head." In medieval times this was transliterated into Latin as *raztaben, rahtaben, razcaben*, and several other corruptions (eventually leading to the name for β Dra – see that star). By Renaissance times the form *Rastaben* had appeared, and this name was erroneously attributed, in its last part, to the Arabic word *thu^(-bān*, "serpent," rather than to *al-tinnīn* (subsequently *thu^(bān* was wrongly regarded as a sci-A name for the constellation Draco). Finally in recent times, the erroneous word, written as "Thuban," was applied as a star name to α Dra.

or Adib (ŭ dēb')
:   From the ind-A name *al-dhi^)b*, "the Wolf," for ζ Dra (originally ζ and η Dra were the ind-A *al-dhi^)bān*, "the Two Wolves"). "Adib" was wrongly transferred to θ Dra in recent times, and still more recently to α Dra.

β Rastaban (räs' tŭ bän') [rŭs' tə băn]
:   Applied with various spellings since medieval times, from the sci-A name *ra^)s al-tinnīn*, "the Serpent's Head," for γ Dra. Transferred to β Dra in recent times (originally Ptolemy had γ Dra on the "head," and β more specifically on the "eye").

or Alwaid (ŭl wä' ĭd) [ăl wād']
:   Applied in recent times from the ind-A asterism name *al-^(awā^)idh*, "the Old Mother Camels," for γ, β, ξ, and ν Dra (today's "Lozenge").

γ Eltanin (ĕl tŭ nēn') [ĕl tā' nĭn]
:   Applied in Renaissance times from the sci-A constellation name *al-tinnīn*, "the Serpent," for Draco.

δ Altais [ăl′ tās]

Applied in recent times from a misreading of the script for the sci-A constellation name *al-tinnīn,* "the Serpent," as it appeared incorrectly in a late sci-A star catalog, as part of the name for δ Dra. (The incorrect word, *al-tais,* is a real word in Arabic meaning "the He-goat," however it was never used by the ind-Arabs in stellar nomenclature. Furthermore, in the same Arabic source, in the name of ε Dra [formed parallel to that of δ Dra], the word was correctly written *al-tinnīn.*)

ι Edasich [ĕ′ də sĭk]

Applied in recent times from its ind-A name *al-dhīkh,* "the Male Hyena."

λ Giausar (jou zŭr′) [jô′ zär]

From the Persian word *jauzahr,* a technical term designating the nodes of the moon's or any planet's orbit. Erroneously applied as a star name to λ Dra in recent times.

μ Alrakis (ŭl rä′ kĭs) [ăl rä′ kĭs]

Applied in recent times from its ind-A name *al-rāqiṣ,* "the Trotting Camel."

ξ Grumium (grū′ mĭ ùm)

A Late Latin word (its correct spelling should be *grunnum*) meaning "snout, or muzzle (especially of a pig)," used in the Medieval Latin *Almagest* in describing this star. (Ptolemy and the sci-Arabs described this star on the serpent's "jawbone.") The word was applied as a star name to ξ Dra in recent times.

EQUULEUS (Equ)

α Kitalpha [kĭ tăl′ fə]

Applied in recent times from the sci-A constellation name *qiṭ′at al-faras,* "the Section of the Horse," for Equuleus. Ptolemy had ῞Ιππου προτομή, "the First Half of the Horse," for the constellation, because it shows only half the figure (as with the constellations Taurus, Pegasus, and the Greek ship Argo).

ERIDANUS (Eri)

α A c h e r n a r (ä′ kĕr när′) [ā′ kər när]
From the sci-A name *ākhir al-nahr*, "the River's End," for θ Eri (see
that star). Transferred to α Eri in Renaissance times, when the constellation Eridanus was extended south to this new terminus.

β C u r s a (kŭr′ sŭ)
Applied in recent times from an abbreviation of the ind-A asterism
name *kursīy al-jauzā³ al-muqaddam*, "the Foremost Footstool of *al-
jauzā³* [today's Orion]," for λ, β, and ψ Eri, and τ Ori. (A "hindmost
footstool" was marked by α, β, γ, and δ Lep.)

γ Z a u r a k (zou′ rŭk) [zô′ răk]
From the Arabic word *zauraq*, "boat," used in an imaginative, non-
classical description of several stars near Eridanus (evidently the stars
of today's Phoenix – see also α Phe). Arbitrarily applied as a star name
to γ Eri in recent times.

ζ Z i b a l (zĭ bäl′)
Applied in recent times from a misreading of the ind-A name *al-ri³āl*,
"the Young of the Ostriches," for the numerous dim stars running
between α Eri and α PsA (or, according to another tradition, those
within the triangle formed by α Phe/α PsA/β Cet).

η A z h a [ŭ′ zə]
Applied in recent times from an abbreviation of the ind-A asterism
name *udḥīy al-naʿām*, "the Ostrich's Nest," for ζ, ϱ^{2,3}, η, and τ$^{1-5}$ Eri,
and ε and π Cet.

θ A c a m a r (ä′ kŭ mär′) [ā′ kə mär]
From its sci-A name *ākhir al-nahr*, "the River's End." Applied in
medieval times, and reapplied in recent times (see α Eri for the differ-
ently-spelled Renaissance application).

or D a l i m (dŭ lēm′)
Applied in recent times from the ind-A name *al-ẓalīm*, "the Ostrich,"
for α Eri (α Eri and α PsA were the ind-Arabs' *al-ẓalīmān*, "the Two

Ostriches"). The sci-Arabs had transferred *al-ẓalīm* to θ Eri, because α Eri was too far south for them to see.

o¹ Beid (bīd)

Applied in recent times from the ind-A name *al-baiḍ*, "the Eggs," for unspecified dim stars around the ind-A ostrich's nest (see η Eri).

o² Keid (kīd)

Applied in recent times from the ind-A name *al-qaiḍ*, "the Egg Shells," for unspecified stars around the ind-A ostrich's nest (see η Eri).

43 (or d) Theemin (tĕ ĕ mēn′) [thē′ mĭn]

The formation of this name begins with one of the words used by Ptolemy to describe this star in the *Almagest*: ἡ καμπή, "the bend [of the river]." This was misread by the sci-Arabs, who transliterated the word as *bhmn*. This, in turn, was transliterated as *beemun* in the Medieval Latin *Almagest*, which was corrupted to *beemin, beemim*, etc. Then in Renaissance times, the derivation of the form *beemim* was erroneously attributed to the Hebrew word *tᵉ)ōmīm*, "twins." Subsequently this erroneous word, written as *Theemim*, then "Theemin," was applied as a star name to any of the various dim stars running from υ¹ to h Eri.

GEMINI (Gem)

α Castor (kŭs′ tŏr) [kăs′ tər]

From its ancient Greek name Κάστωρ, a character in Greek mythology, the twin of Pollux (β Gem). Reapplied in Renaissance times.

As noted under α Cep, Castor's sci-A name (based on ind-A) was *muqaddam al-dhirāʿain*, "the Preceding One of the Two Cubits, or Forearms." This was derived from the ind-A name *al-dhirāʿān* for the two pairs α/β Gem and α/β CMi, where those stars represented either "two cubits" (a cubit is a traditional unit of measure marked by a person's forearm, from elbow to fingertip), or, the "two forearms" of the ind-A asterism *al-asad* (see ε Gem). In addition, the single pair α/β Gem marked the ind-A lunar mansion *al-dhirāʿ*, "the Cubit, or Forearm."

β Pollux (pŏl′ lŭks)
This is the classical Latin form of its ancient Greek name Πολυ-
δεύκης, a character in Greek mythology, the twin of Castor (α Gem).
Reapplied in Renaissance times.

γ Alhena (ŭl hē′ nŭ)
Applied in recent times from the ind-A lunar mansion name *al-han⁽a*,
possibly meaning "the Mark on the Neck of a Camel," for γ and ξ
Gem, or alternatively for γ, ξ, η, μ, and ν Gem.

δ Wasat (wŭ′ sŭt) [wā′ sət]
From the Arabic word *wasaṭ*, "middle," used by a sci-A commentator
who was speculating on the meaning of the ind-A constellation name
al-jauzā⁾. Reference was made to *al-jauzā⁾*'s being in the "middle"
(*wasaṭ*) of the sky (perhaps meaning the celestial equator). The reference was intended to apply to the ind-A figure located in today's
Orion, however it was under the constellation which is today Gemini
that the reference appeared (note the confusion between Orion and
Gemini mentioned under α Ori). Therefore, after transliteration into
Latin, it was in Gemini (to δ) that the word "Wasat" was arbitrarily
applied as a star name in recent times.

ε Mebsuta (mĕb sū′ tŭ)
From an abbreviation of the ind-A asterism name *dhirā⁽ al-asad al-
mabsūṭa*, "the Lion's Outstretched Paw," for one of the two pairs α/β
Gem or α/β CMi. These pairs were assigned to the larger ind-A asterism *al-asad*, but there is no unanimity in the Arabic sources as to
which pair was the "outstretched" or "folded" paw. "Mebsuta" was
arbitrarily applied as a star name to ε Gem in recent times.
In its indigenous location stretching from parts of Gemini all the way
into Virgo, the ind-A *al-asad*, "the Lion," corresponded to what in
other cultures was Leo among the zodiacal signs. Subsequently the sci-
Arabs used the indigenous name for the smaller Greek Lion (Leo).

ζ Mekbuda (mĕk bū′ dŭ)
From an abbreviation of the ind-A asterism name *dhirā⁽ al-asad al-
maqbūḍa*, "the Lion's Folded Paw" (see ε Gem). Arbitrarily applied as
a star name to ζ Gem in recent times.

η P r o p u s (prŏ′ pūs) [prō′ pəs]

Applied in Renaissance times from the Greek word πρόπους, "forward foot," used by Ptolemy in the *Almagest* in describing this star. Earlier charts applied the name to 1 (H) Gem, also in accordance with the *Almagest*.

μ T e j a t (tĕ yät′)

From the Arabic word *tiḫyāt*, said to be the singular form of *al-taḫāyī*, an ind-A name of unknown meaning and disputed identification. In some sources, the plural *al-taḫāyī* is assigned to η, μ, and ν Gem. As originally applied in recent times, η was "Tejat Prior," and μ was "Tejat Posterior."

GRUS (Gru)

α A l N a ' i r (ŭl nä′ ĭr) [ăl nâr′]

Applied in recent times from an abbreviation of its late Arabic name *al-nayyir min dhanab al-ḫūt*, "the Bright One from the Fish's Tail" (where a 16th century Arabic astronomer had extended Ptolemy's Southern Fish [Piscis Austrinus] into what is today Grus). The form Al Na'ir was taken from a wrong transliteration (Al Nāʾir) of the Arabic adjective *al-nayyir*, "the Bright One."

HERCULES (Her)

α R a s a l g e t h i (räs′ ŭl jé′ thē)

Applied with various spellings since medieval times, from its sci-A name *raʾs al-jāthī*, "the Kneeler's Head."

β K o r n e p h o r o s (kŏr né′ fŏ rŏs)

From the Greek word κορυνηφόρος, "club-bearer," mentioned in one Renaissance study as a name for the constellation Hercules. Properly it is of mythological rather than astronomical significance, for Ptolemy's constellation here was "the Kneeler," described without a club. "Kornephoros" was applied as a star name to β Her in recent times.

or R u t i l i c u s (rū tĭ′ lĭ kŭs)
Applied in Renaissance times from the Latin word *titillicus,* "armpit,"
used in the Medieval Latin *Almagest* in describing this star. *Ascella* was
the usual word for "armpit" used in translating from the Arabic (for
example with ζ Sgr), but here the rare word *titillicus* was used. It
became variously misspelled as *rutilicus,* etc., in subsequent manuscripts of the *Almagest* and its derivatives (such as the "Alphonsine
Tables").

ϰ M a r s i c (mŭr′ sĭk) [mär′ sĭk]
Applied in medieval times from the Arabic word *al-marfiq,* "the
elbow," used in the Arabic *Almagest* in describing this star (cf.
λ Oph).

λ M a a s y m (mŭ′ ə sĭm)
From the Arabic word *al-miʿṣam,* "the wrist," used in the Arabic
Almagest in describing o Her. Wrongly transferred to λ Her in
Renaissance times.

ω C u j a m [kū′ yăm]
From the Latin word *caiam,* "club" (in the accusative case), sup-
posedly used in a classical poem alluding to the mythological figure
Hercules, not the constellation (see also β Her). The word was applied
as a star name to ω Her in Renaissance times (but as a name it should
be correctly spelled "Caia," in the nominative case).

HYDRA (Hya)

α A l p h a r d (ŭl fŭrd′) [ăl′ färd]
Applied in medieval times from its ind-A name *al-fard,* "the Solitary
One," descriptive of this star's standing out among the surrounding
dim stars. (See the same word in the plural, with ζ CMa.)

LEO (Leo)

α R e g u l u s (ré′ gù lùs) [rĕ′ gyū ləs]
The formation of this name begins with the ancient Greek name for
this star: Βασιλίσκος, "the (little) King," having obvious origins

among the Sumerians and Babylonians. The Romans, following the Greeks, called the star *stella regia*, "the Royal Star." Later, the Medieval Latin *Almagest*, translating from the Arabic (where in one version the original Greek diminutive had been lost), had *rex*, "the King." The form "Regulus" (again "the [little] King") first appears in 1522 (considerably earlier than in Copernicus' *De revolutionibus orbium coelestium* of 1543, which previous authors have cited as the first use of the name), obviously formed by some Renaissance scholar after the Greek diminutive name. (See also α Aur regarding diminutives.)

β D e n e b o l a [dĕ nĕ' bŏ lə]
Applied with various spellings since medieval times, from its sci-A name *dhanab al-asad*, "the Lion's Tail."

γ A l g i e b a (ŭl jĕ' bŭ) [ăl jē' bə]
Applied in recent times from the ind-A lunar mansion name *al-jabha*, "the Forehead," for ζ, γ, η, and α Leo, associated with the ind-A asterism *al-asad* (see ε Gem).

δ Z o s m a (zös' mŭ)
From the Greek word ζῶσμα, "girdle, or loin cloth." In Renaissance times this word was erroneously said to have been used to describe δ Leo in a medieval Greek text. Correctly, the text has ὀσφῦς, "hip, or lower back" (after Ptolemy). The name "Zosma" (transliterated from the erroneous Greek word) was applied as a star name to δ Leo in recent times.

or D u h r (dŭr)
Applied in recent times from an abbreviation of its sci-A name *ẓahr al-asad*, "the Lion's Back."

ζ A d h a f e r a (ŭ dŭ fē' rŭ) [ə dā' fĕ rə]
From the sci-A name *al-ḍafīra*, "the Lock of Hair," for the Coma Berenices cluster. Wrongly applied to ζ Leo in recent times (due in part to the fact that Ptolemy's "Hair" is discussed under the Lion constellation in the *Almagest*).

θ C h e r t a n (kĕr tän′) [chĕr′ tăn]

    Applied in recent times from one spelling of the ind-A lunar mansion
name *al-khurtān*, "the Two Small Ribs (at or near the breast)," for δ
and θ Leo. The more common and apparently correct spelling of the
lunar mansion name is *al-kharātān*, of unknown meaning.

  or C o x a (kŏk′ sŭ)

    From the Latin term "in coxis," intended as "in the Hips," originally
for θ Leo and the unidentified 21st star of Ptolemy's Lion constellation. The term was used in a Renaissance translation of a late Islamic
star catalog (the latter a Persian source, with many terms retained from
the Arabic *Almagest*). The singular "Coxa," "the Hip," was then
applied as a star name to θ Leo in recent times.

    In comparison, the Medieval Latin *Almagest* consistently used the
word *coxa* to mean "thigh," not "hip," and the word appeared as such
in the constellation Leo in describing the stars ι and σ. Furthermore,
our two stars, θ and the 21st of Leo, were poorly described as on the
pixis vertebri, "socket of the vertebra" (where *vertebrum* was intended
in the same sense as *ancha*, a word properly used for "hip" elsewhere
in the Medieval Latin *Almagest* [for example see θ Aqr]).

λ A l t e r f (ŭl tĕrf′) [ăl′ tǝrf]

    Applied in recent times from the ind-A lunar mansion name *al-ṭarf*,
"the Glance," for ϰ Cnc and λ Leo, associated with the ind-A asterism
al-asad (see ε Gem), as if being located on that lion's eyes.

μ R a s a l a s [rŭ′ sǝ lăs]

    Applied in recent times from an abbreviation of its sci-A name *raʾs al-
asad (al-shamālī)*, "(the Northern [Part] of) the Lion's Head." As
originally applied, μ Leo was "Rasalasad Borealis" (abbreviated in the
first Western astronomical source as "Rasalas. Bor.," for lack of space
in the line), and ε Leo was "Rasalasad Australis" (after the sci-A).

o S u b r a (zŭb′ rŭ) [sū′ brǝ]

    From the ind-A lunar mansion name *al-zubra*, "the Mane, or Shoul-
der," for δ and θ Leo, associated with the ind-A asterism *al-asad* (see
ε Gem; *al-zubra* was an alternative name for *al-kharātān* given under
θ Leo). "Subra" was wrongly applied as a star name to o Leo in recent
times.

LEPUS (Lep)

α Arneb (ŭr′ nĕb) [är′ nĕb]
   Applied in recent times from the sci-A constellation name *al-arnab*,
   "the Hare," for Lepus.

β Nihal (nǐ häl′) [nī′ ăl]
   Applied in recent times from the ind-A asterism name *al-nihāl*, "the
   Camels Beginning to Quench Their Thirst," for α, β, γ, and δ Lep.

LIBRA (Lib)

α Zubenelgenubi (zù bén′ ĕl jĕ nū′ bē) [zū bĕn′ –]
   Applied in Renaissance times from its sci-A name *al-zubānā al-janūbī*,
   "the Southern Claw (of the Scorpion)." The stars of Libra were interpreted by the Babylonians as "the Claws of Scorpius" (thus forming a
   much larger Scorpion), and alternatively (perhaps in a more recent
   stage) they were made an independent constellation, "the Balance."
   Both these conceptions lived on until the Greeks and, partly, the ind-
   Arabs (cf. also β Sco, Graffias). Previous to Greek influences, the ind-
   Arabs used the name *al-zubānayān*, "the Two Claws (of the Scorpion)," for the lunar mansion marked by α and β Lib. (Ind-A *al-
   zubānayān* is obviously related etymologically to Babylonian
   zibānītu, but the latter, also marked by α and β Lib, meant "the
   Balance.") Our modern name is occasionally seen abbreviated as
   "Zuben'ubi."

β Zubeneschamali (zù bén′ ĕ shŭ mä′ lē) [zū bĕn′ ĕ shə mä′ lē]
   Applied in Renaissance times from its sci-A name *al-zubānā al-
   shamālī*, "the Northern Claw (of the Scorpion)," (see α Lib).

LYRA (Lyr)

α Vega (vé′ gŭ) [vē′ gə]
   Applied in medieval times from an abbreviation of its ind-A name *al-
   nasr al-wāqiʿ*, "the Swooping Eagle (or Vulture)," alternatively used as

an asterism name for α, ε, and ζ Lyr (see also α Aql). "Vega" is one of the oldest Arabic star names applied in the West, from the end of the 10th century A.D.

β Sheliak (shĕl yäk') [shĕl' yăk]
    Applied in recent times from the sci-A constellation name *al-salbāq*, "the Harp," for Lyra. *salbāq* was a Greek loanword in Arabia (from σαμβύχη, a kind of harp), and was one of three names of musical instruments used by the sci-Arabs for Lyra. Still other names were *al-sulaḥfāt* (see γ Lyr), and *al-lūrā* (a transliteration of Ptolemy's Λύρα, that became Latinized in medieval times as *allore*).

γ Sulafat (sù lŭ fät') [sū' lə făt]
    Applied in recent times from the sci-A constellation name *al-sulaḥfāt*, "the Tortoise," for Lyra, corresponding to the Greek image of a tortoise shell for the Lyre's soundboard (see also β Lyr and π¹ Cyg).

OPHIUCHUS (Oph)

α Rasalhague (räs' ŭl hŭ wé') [räs' ăl hā' gwē]
    Applied with various spellings since medieval times, from its sci-A name *raʾs al-ḥawwāʾ*, "the Head of the Serpent Collector."

β Cebalrai (sĕb' ŭl rä' ē) [sĕ' băl rä' ē]
    Applied in recent times from its ind-A name *kalb al-rāʿī*, "the Shepherd's Dog" (in other traditions α Her and 28/29 Cep are given the same name). This dog, together with the shepherd (marked by α Oph), and the two lines of stars enclosing the Pasture (see β CrB), and all the dim stellar sheep in that portion of the sky (*al-ghanam* or *al-aghnām* for the sheep), seem to form a complete group of ind-A figures. (Compare to a different group discussed under γ Cep.)

δ Yed Prior (yĕd – prē' ŏr) [– prī' ər]
ε Yed Posterior (yĕd – pŏs tē' rĭ ŏr)
    Originally from the Arabic word *al-yad*, "the hand," used in the Arabic *Almagest* in describing these stars. *yed* was applied as a star name to δ Oph in medieval times. Then in recent times, this applica-

tion was extended to include both δ and ε, with the addition of the Latin distinctions for "foremost" and "hindmost," respectively.

η Sabik (sä' bĭk) [sä' bĭk]

Applied in recent times from its ind-A name *al-sābiq*, of uncertain meaning or connection. The name might mean literally "the Preceding One," or "the One Arriving First in a Race." Other sources give the name in the dualis form (for the two stars ζ and η Oph), and in the plural form (for more than two stars).

λ Marfik (mŭr' fĭk) [mär' fĭk]

Applied since medieval times from the Arabic word *al-marfiq*, "the elbow," used in the Arabic *Almagest* in describing this star (cf. ϰ Her). While the medieval transliteration was *marsic* (with a confusion of *f* and the long-shaped *s*), Marfik is an improved spelling of recent times.

ORION (Ori)

α Betelgeuse (bĕt' ĕl jou zé') [bĕ' təl jūz, bē' təl jūs, etc.]

Applied with varied spellings since medieval times, from its ind-A name *yad al-jauzā*, "the Hand of *al-jauzā*." (The first medieval transliteration into Latin was *bedalgeuze*, mistaking the initial Arabic letter as "*b*" instead of "*y*." In Renaissance times this corruption was erroneously attributed to the assumed Arabic word "*bāṭ*" [properly *ibṭ*], for "the Armpit" of *al-jauzā*, giving rise to the corruption "Betelgeuse" seen today. Thus by Renaissance times both the "*y*" and "*d*" of the first part of the ind-A name had become corrupted.)

The ind-A asterism *al-jauzā* was a feminine figure represented in the stars of what is today Orion. The origin of its name is unclear (as with nearly all of the ancient group of ind-A names). The root *jwz* can mean "middle," and the word *al-jauzā* is structured as a feminine adjective, thus *al-jauzā* may mean "the female one, having something about her related to the middle." There have been different speculations as to what the "middle" reference, if that is what it is, could be (for one example, see δ Gem).

The ind-Arabs' *al-jauzā* corresponded to what in other cultures was Gemini among the zodiacal signs. The sci-Arabs subsequently used the indigenous name for both the Greek Orion and Twins, leading to some confusion in star names between the two constellations. An alternative sci-A name for Orion was *al-jabbār*, "the Giant;" and for Gemini, *al-taw'amān*, "the Twins."

β R i g e l (rĭjl) [rī′ jəl]
From an abbreviation of its ind-A (and sci-A) name *rijl al-jauzā'*, "the Foot of *al-jauzā'*." "Rigel" is one of the oldest Arabic star names applied in the West, from the end of the 10th century A.D.

γ B e l l a t r i x (bĕl lä′ trēks) [bĕl lā′ trĭks]
A Latin name meaning "the Female Warrior," given to α Aur in a medieval astrological text (the reason behind this initial application is not fully understood, and awaits further findings in corresponding Arabic texts). In late medieval times, *bellatrix* was transferred to γ Ori, perhaps by association with the name *Bellator*, "the Male Warrior," which was used for Orion in other astrological texts (presumably *Bellator* was a tentative translation of Orion's sci-A name *al-jabbār*).

δ M i n t a k a (mĭn′ tŭ kŭ)
Applied in recent times from an abbreviation of the ind-A (and sci-A) asterism name *minṭaqat al-jauzā'*, "the Belt (or Girdle) of *al-jauzā'*," for δ, ε, and ζ Ori (compare to alternative names given under ε and ζ Ori).

ε A l n i l a m (ŭl nĭ läm′) [ăl nī′ lăm]
Applied in recent times from the ind-A asterism name *al-niẓām*, "the String of Pearls," for δ, ε, and ζ Ori.

ζ A l n i t a k (ŭl nĭ täk′) [ăl nī′ tăk]
Applied in recent times from an abbreviation of the ind-A asterism name *niṭāq al-jauzā'*, "the Belt (or Girdle) of *al-jauzā'*," for δ, ε, and ζ Ori.

ϰ S a i p h (sāf)
From an abbreviation of the sci-A asterism name *saif al-jabbār*, "the Giant's Sword," for η, c, θ, and ι Ori. Wrongly applied as a star name to ϰ Ori in recent times.

λ Meissa (mā sä′) [mī′ sə]
From the ind-A name *al-maisān*, for either γ or ξ Gem. The meaning of the name is uncertain. It may mean "the Sparkling One," referring to a star, or "the Proudly Marching One," referring to a person. Together, γ and ξ Gem composed the ind-A 6th lunar mansion *al-han'a* (see γ Gem). The nearby stars λ, φ¹, and φ² Ori composed the adjacent 5th lunar mansion *al-haq'a* (see Heka below). Due to confusion between these two parent lunar mansions in a late Arabic source, "Meissa" was wrongly applied to λ Ori (rather than to γ or ξ Gem) in recent times. "Meissa" (without n at the end) corresponds to ind-A *al-maisā'*, a variant form of *al-maisān*.

or Heka (hĕ′ kŭ)
Applied in recent times from the ind-A lunar mansion name *al-haq'a*, "the Circle of Hair on (the side, neck, or foot of) a Horse," for λ, φ¹, and φ² Ori.

PAVO (Pav)

α Peacock (pē′ kŏk)
Applied as a star name to α Pav in recent times, being the English translation of the constellation name Pavo.

PEGASUS (Peg)

α Markab (mŭr′ kŭb) [mär′ kăb]
From an abbreviation of the sci-A name *mankib al-faras*, "the Horse's Shoulder," for β Peg. Wrongly transferred to α Peg in late medieval times.

β Scheat [shē′ ăt]
From the Arabic word *al-sāq*, "the shin," used in the Arabic *Almagest* in describing δ Aqr (see that star). Wrongly transferred as a star name to β Peg in late medieval times.

γ Algenib (ŭl jĕ′ nĭb) [ăl jē′ nĭb]
From the sci-A name *al-janb*, "the Side," for α Per (see there). Wrongly transferred to γ Peg in Renaissance times.

ε Enif (ĕ′ nĭf)
    Applied in medieval times, evidently from the Arabic word *anf*,
    "nose" (Ptolemy had described this star on the horse's "muzzle").
    However, sci-A sources do not mention *anf* for ε Peg, only other
    terms, hence the ultimate origin of "Enif" remains uncertain.

ζ Homam (hŏ mäm′) [hō′ măm]
    Applied in recent times from an abbreviation of the ind-A name *sa ͑d
    al-humām*, for ζ and ξ Peg. A possible meaning for the name is "the
    Lucky (Stars) of the Hero," but the exact historical connections are
    unknown (see α Aqr).

η Matar (mŭ′ tŭr) [mä′ tär]
    Applied in recent times from an abbreviation of the ind-A name *sa ͑d
    maṭar* for η and o Peg. Its meaning is unknown (see α Aqr). *maṭar* has
    been translated as "rain," and as a common noun in Arabic, *al-maṭar*
    does mean "the rain." However, lack of the definitive article *al-* for the
    star points to a use of the word in a sense different from the common
    noun.

θ Biham (bĭ häm′) [bī′ ăm]
    Applied in recent times from an abbreviation of the ind-A name *sa ͑d
    al-bihām*, for θ and ν Peg. A possible meaning for the name is "the
    Lucky (Stars) of the Young Beasts [lambs, kids, and the like]," but the
    exact historical connections are unknown (see α Aqr).

μ Sadalbari (säd′ ŭl bä′ rĭ) [sŭd′ ăl bä′ rē]
    Applied in recent times from the ind-A name *sa ͑d al-bāri ͑*, for λ and
    μ Peg. A possible meaning for the name is "the Lucky (Stars) of the
    Excelling One," but the exact historical connections are unknown
    (see α Aqr).

PERSEUS (Per)

α Mirfak (mĭr′ fŭk) [mĭr′ făk]
    Applied in recent times from an abbreviation of its ind-A name *mirfaq
    al-thurayyā*, "the Elbow of the Pleiades," for its location in the ind-A
    asterism here (see β Cas).

or Algenib (ŭl jĕ′ nĭb) [ăl jē′ nĭb]

Applied with various spellings since medieval times, from the Arabic word *al-janb*, "the side, or flank," used in the Arabic *Almagest* in describing this star. Renaissance scholars explained the word erroneously as from the Arabic *al-jānib* (instead of the correct *al-janb*, but of the same meaning), transliterated as "Algenib," and this last form afterwards was adopted by Western astronomers.

β Algol (ŭl göl′) [ăl′ gŏl]

Applied in medieval times from an abbreviation of its sci-A name *ra's al-ghūl* "the Demon's Head" (for Ptolemy's Gorgon-head). "Algol" is one of the oldest Arabic star names applied in the West, from the end of the 10th century A.D.

ξ Menkib (mĕn′ kĭb)

Applied in recent times from an abbreviation of its ind-A name *mankib al-thurayyā*, "the Shoulder of the Pleiades," for its location in the ind-A asterism here (see β Cas).

o Atik (ä′ tĭk) [ā′ tĭk]

Applied in recent times (also to ζ Per), from an abbreviation of the ind-A name *'ātiq al-thurayyā*, "the Collarbone of the Pleiades," for o and ζ Per, after their location in the ind-A asterism here (see β Cas).

PHOENIX (Phe)

α Ankaa (ŭn kä′) [ăn′ kə]

Applied in recent times from the modern Arabic constellation name *al-'anqā'* (a fabulous bird) for Phoenix.

or Nair al Zaurak (nä′ ĭr – ŭl zou′ rŭk) [nâr – ăl zô′ răk]

Applied in recent times from its late Arabic name *nayyir al-zauraq*, "the Bright One of the Boat." This was based on an imaginitive, non-classical description of several stars near Eridanus, including α Phe (see also γ Eri). For the wrong transliteration of the Arabic word *nayyir* cf. above, ζ Cen and α Gru.

PISCES (Psc)

α Alrescha (ŭl rĕ shä') [ăl rē' shə]
From the ind-A lunar mansion name *al-rishā⁾*, "the Cord," for β
And. Wrongly transferred to α Psc in recent times.
According to a sci-A source, the ind-A *al-rishā⁾* included, in a larger
sense, two curving lines of mostly dim stars in Andromeda and Pisces,
meant to attach to the ind-A *al-dalw* "the Well Bucket" (marked by
today's Square of Pegasus). It is also possible that this "cord" was a
remnant of the cord joining the two fish of the older Babylonian
zodiac. Indeed, these stars of the cord were alternatively known by the
ind-Arabs as *al-ḥūt*, "the Fish" (a single fish), corresponding to what
in other cultures was Pisces among the zodiacal signs (and according to
this tradition, the Fish's brightest star, β And, was *baṭn al-ḥūt*, "the
Fish's Belly").

PISCIS AUSTRINUS (PsA)

α Fomalhaut (fŏm' ŭl hout') [fō' mă lôt, fō' mă lō]
Applied with various spellings since medieval times, from an abbreviation of its sci-A name *fam al-ḥūt al-janūbī*, "the Mouth of the Southern Fish."

PUPPIS (Pup)

ζ Naos (nous)
Applied in recent times from the word Ναῦς, Greek for the "Ship,"
used in one Renaissance discussion as a name for the ancient constellation Argo (since divided into Puppis, Carina, Vela, and Pyxis).

ϱ Tureis (tù rās')
From the Arabic word *al-turais*, "the Little Shield," used in the Arabic
Almagest in describing several stars (ϱ Pup not among them) in the
constellation Argo (see ι Car). Wrongly applied as a star name to
ϱ Pup in recent times.

SAGITTA (Sge)

α Sham [shăm]
Applied in recent times from the sci-A constellation name *al-sahm*,
"the Arrow," for Sagitta.

SAGITTARIUS (Sgr)

α Rukbat (rŭk′ bŭt) [rŭk′ băt]
Applied in recent times from an abbreviation of its sci-A name *rukbat
al-rāmī*, "the Archer's Knee."

or Alrami (ŭl rä′ mē)
Applied in recent times from the sci-A constellation name *al-rāmī*,
"the Archer," for Sagittarius.

β Arkab [är′ kăb]
Applied in recent times from an abbreviation of its sci-A name *ʿurqūb
al-rāmī*, "the Archer's Achilles Tendon."

γ Alnasl (ŭl nŭsl′) [ăl nāzl′]
Applied in recent times from its late Arabic name *al-naṣl*, "the Point,"
in turn an abbreviation of the Arabic *Almagest*'s *naṣl al-sahm*, "the
Point of the Arrow."

or Nushaba (nù shä′ bŭ)
Applied in recent times from its late Arabic name *zujj al-nushshāba*,
"the [Iron] Point of the [Wooden] Arrow," cited in a Renaissance
discussion of Sagittarius.

δ Kaus Media (kous′ – mé′ dĭ ŭ) [kôs′ – mē′ dĭ ə]
ε Kaus Australis (kous′ – ous trä′ lĭs) [kôs′ – ôs trä′ lĭs]
Arbitrarily applied in recent times, together with the Latin distinctions
of "middle," "southern," and "northern" (for λ Sgr), from the ind-A
(and later sci-A) constellation name *al-qaus*, "the Bow."
For the ind-Arabs, *al-qaus* was marked by the curved line of stars ξ²,
o, π, d, ϱ, and υ Sgr (corresponding to what in other cultures was
Sagittarius among the zodiacal signs). The sci-Arabs used the indigenous name for the Greeks' Archer, alternatively translated as *al-rāmī*.

ζ Ascella (ŭ sĕl′ lŭ)
   A Latin word meaning "armpit," used in the Medieval Latin *Almagest*
   in describing this star. Applied as a star name to ζ Sgr in recent times.

λ Kaus Borealis (kous′ – bŏ′ rĕ ä′ lĭs) [kôs – bŏ′ rĕ ä′ lĭs]
   (See δ and ε Sgr.)

σ Nunki (nŭn′ kē′)
   Applied in recent times from some Babylonian name NUN^{ki} (as written in Sumerian ideograms), an untranslated proper name. The name
   was probably for a star, or stars, in today's Vela, Puppis, or Carina,
   and may have been the name of α Car. Furthermore, the Babylonians
   regarded NUN^{ki} as the celestial counterpart of their sacred city Eridu,
   city of the god Ea, on the Euphrates River.

SCORPIUS (Sco)

α Antares (ŭnt ä′ rés) [ăn tā′ rēz]
   From its ancient Greek name Ἀντάρης, "Like Ares," likening the red
   color of this star to its planetary namesake (the Roman Mars). Reapplied in Renaissance times. (The Greek preposition ἀντί can mean
   "like" or "in place of," as in the present case.)

β Acrab (ŭk′ rŭb) [ək răb′]
   Applied in recent times from the ind-A (and sci-A) constellation name
   al-ʿaqrab, "the Scorpion." *al-ʿaqrab* was the ind-A name for the
   zodiacal constellation of Scorpius, coinciding with the ancient Greek
   (and modern) constellation both in name and location. (See also
   θ Sco.)

   or Graffias (grŭf′ fĭ äs) [grăf′ fĭ äs]
   A Latinized Romance word in the accusative case meaning "claws,"
   used in a Medieval Latin translation of Ptolemy's *Tetrabiblos* (translated through Old Spanish and Arabic). Reference in the *Tetrabiblos*
   was intended for the stars of Libra (understood, in classical manner, as
   the "claws" of Scorpius), however in Renaissance times the word was
   arbitrarily applied as a star name to ξ Sco. Then in recent times, the

name was transferred to β Sco (where "Graffia" would be the correct singular, nominative case).

δ D s c h u b b a (jŭb′ bŭ)
Applied in recent times from an abbreviation of the sci-A name *jabhat al-ʿaqrab*, "the Scorpion's Forehead," for β, δ, and π Sco.

θ G i r t a b (gĭr′ täb′, gĭr′ tŭb′)
Applied in recent times from the Sumerian constellation name GÍR.TAB, "the Scorpion," for Scorpius. This was translated by the Babylonians as *aqrabu*, which in turn was somehow transmitted to Arabia to become the ind-A (and sci-A) *al-ʿaqrab* (see β Sco).

or S a r g a s (sŭr′ gŭs′) [sär′ gäs]
Applied in recent times from some Sumerian name ŠAR.GAZ, for one of a pair of stars, the other one being ŠAR.UR, the names for two weapons of the god Marduk (approximately to be translated as "the Great Smasher" and "the Great Beam, or Wing"), designating υ and λ Sco, respectively.

λ S h a u l a (shou′ lŭ) [shô′ lə]
Applied in recent times from the ind-A lunar mansion name *al-shaula*, "the Scorpion's Stinger," for λ and υ Sco, part of the ind-A Scorpion (see β Sco).

σ A l N i y a t (ŭl nĭ yät′) [ăl nī′ yăt]
Applied in recent times from the ind-A name *al-niyāṭ*, "the Arteries," for σ and τ Sco. α Sco, between the arteries, was the ind-A *qalb al-ʿaqrab*, "the Scorpion's Heart" (all were part of the ind-A Scorpion – see β Sco).

υ L e s a t h (lĕs′ ŭt)
The formation of this name begins with the Greek term (νεφελοειδὴς) συστροφή, "(foggy) conglomeration," used by Ptolemy in the *Tetrabiblos* in describing several nebulous sky objects (in the present case, probably the open cluster M 7). Ptolemy's term was translated by the sci-Arabs as *al-laṭkha*, "the Spot." This word in turn became transliterated and corrupted in Medieval Latin to *alascha*, which was used in astrological texts in connection with Scorpius' tail. In Renaissance times, the derivation of *alascha* was erroneously attributed to the

Arabic word *lasʿa*, "sting, or bite (of a poisonous animal)," rather than to *al-laṭkha*. Subsequently the erroneous word, written as "Lesath," was applied as a star name to υ Sco (and in recent times, also to λ Sco).

SERPENS (Ser)

α Unukalhai (ů′ nůk ŭl hī′) [yū′ nək ăl hā′ ē]
Applied in recent times from its sci-A name *ʿunuq al-ḥayya*, "the Serpent's Neck."

θ Alya (ŭl′ yŭ)
From the Arabic word *alya*, naming the fatty tail of a breed of Eastern sheep. In Renaissance times this word was erroneously proposed as the origin of *Alioth*, a Medieval Latin name for ε UMa. The erroneous "Alya" was applied as a star name to θ Ser in recent times. (See ε UMa for the correct derivation of *Alioth*.)

TAURUS (Tau)

α Aldebaran (ŭl′ dĕ bŭ rän′) [ăl dĕ′ bə răn]
Applied in medieval times from its ind-A name *al-dabarān*, possibly meaning "the Follower," alternatively used as the lunar mansion name for all the Hyades (or again for α Tau alone). The name is thought to refer to this star's following the Pleiades across the sky, or to the Hyades (or α Tau) coming after the Pleiades as a lunar mansion. "Aldebaran" is one of the oldest Arabic star names applied in the West, from the end of the 10th century A.D.

β Elnath (ĕl nŭt′) [ĕl′ năth]
From the ind-A name *al-naṭḥ*, "the Butting (with the horns)," an alternative name for the 1st lunar mansion *al-sharaṭān* consisting of β and γ Ari (see β Ari). Some late Arabic authors also applied *al-naṭḥ* as a star name to α Ari. In recent times the name, as "Nath," was wrongly transferred to β Tau, and subsequently it was completed to "Elnath" (adding the Arabic definite article).

ε Ain (ān)

 From an abbreviation of the sci-A name ʿain al-thaur, "the Bull's Eye," for α Tau. Applied to ε Tau in recent times (Ptolemy had α Tau on the southern eye, and ε Tau on the northern eye).

η Alcyone (ŭl kĭ′ ŏ né) [ăl sī′ ō nē]
16 Celaeno (kĕ lĭ′ nö) [sĕ lē′ nō]
17 Electra (é lĕk′ trŭ)
19 Taygeta (tä ĭ′ gĕ tŭ) [tā ĭ′ jē tə]
20 Maia (mä′ yŭ) [mä′ yə]
21 Sterope (stĕ′ rŏ pé) or Asterope
23 Merope (mĕ′ rŏ pé)
27 Atlas (ŭt′ lŭs) [ăt′ ləs]
28 Pleione (plé ĭ′ ŏ né) [plē′ yŏ nē]

 These names were individually applied in Renaissance times from a family of characters in Greek mythology: Atlas, Pleione, and their seven daughters, the Pleiades.

TRIANGULUM (Tri)

α Mothallah (mŏ thŭl′ lŭ)

 Applied in recent times from the sci-A constellation name al-muthallath, "the Triangle," for Triangulum.

TRIANGULUM AUSTRALE (TrA)

α Atria (ä′ trĭ ŭ) [ä′ trĭ ə]

 Applied in recent times and obviously coined from its Greek letter designation alpha, plus the constellation name Triangulum Australe.

URSA MAJOR (UMa)

α Dubhe (dŭb′ bə) [dŭ′ bē]

 Applied in medieval times from the sci-A constellation name al-dubb, "the Bear," for Ursa Major. The spelling in the oldest sources was *edubh* (corresponding to the Arabic noun with the article *al* prefixed as *e*). Afterwards, by corruption, the initial *e* was transferred to the end of the word: *dubhe.*

β Merak (mĕ räk′) [mē′ răk]
Applied in recent times from an abbreviation of its sci-A name *marāqq al-dubb al-akbar,* "the Flank (or Groin) of the Greater Bear."

γ Phecda [fĕk′ də]
Applied in recent times from an abbreviation of its sci-A name *fakhidh al-dubb al-akbar,* "the Thigh of the Greater Bear."

δ Megrez (mĕg′ rĕz) [mē′ grĕz]
Applied in recent times from an abbreviation of its sci-A name *maghriz al-dubb al-akbar,* "the Root (of the tail) of the Greater Bear."

ε Alioth (ŭl yöt′) [ă′ lĭ ŏth]
Ultimately from its ind-A name *al-jaun,* "the Black Horse, or Bull," which became corrupted even in Arabic sources (where some of the corruptions carried other meanings). For example, in the Arabic manuscript of the *Almagest* that was translated into Latin in 1175 A.D., this name of ε UMa had apparently been miswritten as *al-jauza* or *al-jauzāʾ* (the latter being identical to the sci-A name for Orion and Gemini – see α Ori). Whatever was read in this Arabic manuscript was transliterated into Latin as *alioze.* In subsequent Latin copies, this name was further corrupted to *aliore, Alcor, Alioth,* etc. Since late medieval times, "Alioth" became the preferred name for ε UMa. (See also 80 UMa and θ Ser.)

ζ Mizar [mī′ zär]
Ultimately from the Arabic word *al-marāqq,* "the Groin," used in the Arabic *Almagest* in describing β UMa. Correct transliterations of *al-marāqq* in the Medieval Latin *Almagest* were *mirac* and *mirach.* However, Renaissance scholars confused these words with the Arabic word *miʾzar,* which had equally been transliterated and corrupted in Latin as *mirac, mirach,* etc. (see β And). Subsequently the mistaken *miʾzar,* written as "Mizar," was applied as a star name to β UMa, and it became transferred to ζ UMa in late Renaissance times.

η A l k a i d (ŭl kä′ ĭd) [ăl kād′]

> Applied with various spellings since medieval times, from its ind-A name *al-qā'id*, "the Leader" (probably as leader of the Daughters of the Bier – see Benetnasch below).

or B e n e t n a s c h (bĕ nét′ näsh′) [bĕ′ nĕt năsh]

> Applied with various spellings since medieval times, from the ind-A asterism name *banāt na'sh*, for α, β, γ, δ, ε, ζ, and η UMa (the familiar figure of the Dipper, Plough, or Wain). *banāt na'sh* is among the most ancient of ind-A names, and its original meaning or significance is unknown. As a common noun in Arabic, *al-na'sh* means "the bier," therefore later sci-A and Western authors were variously inclined to regard the quadrangle formed by α, β, γ, and δ as a bier. Similarly, *al-banāt* means "the daughters," and the three stars ε, ζ, and η have been regarded as "daughters of the bier" (with η UMa, in another ind-A tradition, as their "leader" – see Alkaid above).

ι T a l i t h a (tä′ lĭ thŭ) [tā′ lĭ thə]

> Applied in recent times from an abbreviation of the ind-A name *al-qafza al-thālitha*, "the Third Leap," for ι and ϰ UMa. This pair, with the pairs λ/μ and ν/ξ UMa, composed the ind-A asterism *qafazāt al-ẓibā'*, "the Leaps of the Gazelles," imagined as the tracks left by those animals.
>
> In some later tradition, *qafazāt al-ẓibā'* and adjacent asterisms were associated in a fable, where several gazelles leaped away from the Lion into a pond (*al-ḥauḍ*, marked by τ, h, υ, φ, θ, e, and f UMa), leaving their tracks behind.

λ T a n i a B o r e a l i s (tä′ nĭ yŭ – bŏ′ rĕ ä′ lĭs) [tān′ yə – bŏ′ rē ă′ lĭs]
μ T a n i a A u s t r a l i s (tä′ nĭ yŭ – ous trä′ lĭs) [tān′ yə – ôs trä′ lĭs]

> Applied in recent times from their ind-A name *al-qafza al-thāniya*, "the Second Leap" (see ι UMa), and the Latin distinctions of "northern" and "southern," respectively.

ν A l u l a B o r e a l i s (ŭl ū′ lä – bŏ′ rĕ ä′ lĭs) [ăl ū′ lə – bŏ′ rē ă′ lĭs]
ξ A l u l a A u s t r a l i s (ŭl ū′ lä – ous trä′ lĭs) [ăl ū′ lə – ôs trä′ lĭs]

> Applied in recent times from their ind-A name *al-qafza al-ūlā*, "the First Leap" (see ι UMa), and the Latin distinctions of "northern" and "southern," respectively.

o Muscida (mŭs′ sĭ dŭ) [mū′ sĭ də]
A Latin word meaning "muzzle," used in the Medieval Latin *Almagest* in describing this star. Applied as a star name to o UMa in recent times.

80 Alcor [ăl′ kôr]
Ultimately from the ind-A name *al-jaun*, for ε UMa (see that star), which was transliterated and eventually corrupted in Latin to "Alcor." The name was wrongly transferred to 80 UMa in Renaissance times. (The most common ind-A name for 80 UMa was *al-suhā*, "the Neglected One," as likely to be overlooked beside the brighter star ζ UMa.)

URSA MINOR (UMi)

α Polaris (pŏ lä′ rĭs) [pō lâ′ rĭs]
A Latin adjective meaning "of the pole" (related to the noun *polus*, "the pole"). Applied as a star name to α UMi in Renaissance times, it reflects this star's temporary proximity to the north celestial pole.

or Alrucaba (ŭl rŭ′ kŭ bŭ)
From the sci-A name *al-rukba*, "the Knee (of the Greater Bear)," for θ UMa. Wrongly transferred to α UMi in late medieval times.

β Kochab [kō′ kăb]
Applied to β UMi in Renaissance times and of uncertain derivation. It is probably from one of the names *Alrucaba, Rucaba*, etc., that were first applied to θ UMa, then to α UMi (see that star), in medieval times. However, the name may also be from the Arabic word *kaukab*, or the Hebrew word *kōkhābh*, both meaning "star."

γ Pherkad (fĕr′ kŭd)
Applied in recent times from the singular form of the ind-A asterism name *al-farqadān*, "the Two Calves," for β and γ UMi.

δ Yildun (yĭl dŭn′)
From the Turkish word *yıldız*, "star." In Renaissance times this common noun was erroneously said to be a Turkish name for the Pole Star (α UMi). Misspelled as "Yildun," the word was arbitrarily applied as a star name to δ UMi in recent times.

VELA (Vel)

γ Regor [rē' gôr]

Applied in recent times. The true story of the formation of this name (together with Dnoces for ι UMa and Navi for γ Cas) has been told by E. C. Krupp in Sky & Telescope, October 1994, pp.63–65: the three "names" were formed, as a jest, by V. I. Grissom and T. Jenzano and added to a list of navigational stars for astronauts. Regor is here the reverse of the first name of Roger B. Chaffee, one of the three astronauts who afterwards died in the Apollo accident of January 27, 1967. While it is now often assumed that the three "names" thus constructed are meant to honor the three astronauts who lost their lives in that accident, these names had already been created before this sad event.

χ Markeb (mŭr' kĕb)

Evidently from the Arabic word *markab*, "a ship or any vehicle," presumably standing for the Greek constellation name Argo. However, sci-A sources do not mention *markab* for Argo, only *al-safīna* ("the Ship"), hence the ultimate origin of the name remains a mystery. "Markeb" was applied as a star name to ρ Pup in medieval times, to k Pup in Renaissance times, and finally to χ Vel in recent times when Argo was divided into its four modern constellations.

λ Suhail (sū hāl')

Applied in recent times and representing an abbreviation of any of several composite ind-A names (for example *suhail al-wazn, suhail al-muḫlif*) that sci-A authors variously attributed to brighter stars in the region of *suhail* (see α Car). λ Vel was among these brighter stars. Some of the composite names may have been authentic ind-A names for far-southern stars, with their true identities unknown to the more northern sci-Arabs, while others of them were surely the creations of ind-A poets.

VIRGO (Vir)

α Spica (spē' kŭ) [spī' kə]

Its ancient Roman name meaning "the Ear of Grain," after the star's Greek name Στάχυς of identical meaning (in turn having obvious

origins among the Babylonians and Sumerians). Reapplied in Renaissance times.

or A z i m e c h (ŭ sĭ mék´)
Applied in medieval times from an abbreviation of its ind-A name *al-simāk al-a ͑zal*, "the Unarmed *simāk*" (as opposed to "the Lance-bearing *simāk*" for α Boo [see η Boo]).

β Z a v i j a v a (zä´ vĭ yŭ vä´) [ză´ vĭ jä´ və]
From a contraction of the ind-A name *zāwiyat al-͑awwā ͗*, "the Angle
of *al-͑awwā ͗*," for γ Vir. Wrongly transferred to β Vir in recent times.
The stars β, η, γ, δ, and ε Vir marked the ind-A lunar mansion *al-
͑awwā ͗* (with γ in its "angle"), but the meaning of *al-͑awwā ͗* is unknown.

γ P o r r i m a (pŏr´ rĭ mŭ)
The name of a Roman goddess, one of several mythological names
mentioned in connection with the constellation Virgo. Applied as a
star name to γ Vir in recent times.

ε V i n d e m i a t r i x (vēn dé´ mĭ ä´ trĭks) [vĭn dē´ mĭ ä´ trĭks]
The formation of this name begins with the classical Greek name for
this star: Προτρυγητήρ, "the Grape Gatherer" (where this star's
heliacal rising signalled the onset of the vintage). The Romans translated this name as *vindemiator* and other similar masculine forms.
Similarly, after a somewhat erroneous translation of the Greek name
by the sci-Arabs (see "Almuredin" below), the Medieval Latin translation in the *Almagest* was *precedens vindemiatorem*, "the One Preceding the Grape Gatherer," modified in some editions of the Alphonsine
Tables into *precedens vindemitorem*. The exact origin of the current,
feminine "Vindemiatrix" from any of these forms is uncertain, but it
first appeared for ε Vir in Renaissance times.

or A l m u r e d i n [ăl myū´ rə dĭn]
The formation of this name begins with Ptolemy's name for this star in
the *Tetrabiblos*: Προτρυγητήρ, "the Grape Gatherer." This was translated literally into Arabic as *al-mutaqaddim li-l-qaṭṭāf*, "the One Preceding the Grape Gatherer." This in turn was misread and mistranslit-
erated in one Medieval Latin version of the *Tetrabiblos* as *almucedeme
alacaf*. With subsequent corruption, erroneous Renaissance "correc-
tion," and abbreviation, the initial portion of this Latin name, written
as "Almuredin," was retained as a name for ε Vir.

η Zaniah (zä′ nĭ yŭ) [zān′ yə]

>   From an abbreviation of the ind-A name *zāwiyat al-ʿawwāʾ*, "the
>   Angle of *al-ʿawwāʾ*," for γ Vir (see β Vir). Wrongly transferred to
>   η Vir in recent times.

ι Syrma (sĭr′ mŭ)

>   From the Greek word σύρμα, "train of a dress," used by Ptolemy in
>   the *Almagest* in describing this star. The word was applied as a star
>   name to ι Vir in recent times.

Appendix

The Bečvář Names

Since Antonín Bečvář's *Atlas Coeli*, Prague 1951, there have appeared 14 strange star names, 12 of which have so far escaped all efforts to explain them. These names are:

Achird	η	Cas	Kraz	β	Crv
Arich	γ	Vir	Ksora	δ	Cas
Haris	γ	Boo	Kuma	ν	Dra
Hassaleh	ι	Aur	Reda	γ	Aql
Hatysa	ι	Ori	Sarin	δ	Her
Heze	ζ	Vir	Segin	ε	Cas
Kaffa	δ	UMa	Tyl	ε	Dra

After Bečvář (1951), these names can be found in various astronomical publications of all sorts. Only two of them allow a reasonable explanation: Haris may be from Arabic *ḥāris al-shamāl*, "Guardian of the North", translating in turn the Greek names of either α Boo, 'Αρκτοῦρος, or the constellation Bootes, 'Αρκτοφύλαξ. Segin may be a written variant of Seginus (see above, γ Boo). The remaining 12 names bear no relation whatsoever to any of the languages basically involved in the formation of our star names – Greek, Arabic and Latin.

For the past fifteen years P. Kunitzsch has tried many ways of obtaining information about Bečvář's methods of work and the sources he might have used in preparing the *Atlas Coeli*. He also contacted a number of astronomers in the Czech Republic and in Slovakia in order perhaps to find some former collaborators of Bečvář who might be of some help in this matter, but all without success. Since A. Bečvář died in 1965, the chance of tracing the background of these names is continuously diminishing. (The explanations given in the booklet of Z. Denk and O. Hlad, *Hvězdy s arabskými názvi (Stars with Arabic Names)*, issued by the Prague Planetarium, 1996, are completely arbitrary and cannot be traced in any of the original Greek, Arabic and Latin sources.)

(These names are not entered in the following Index of names)

Index

Acamar θ Eri
Achernar α Eri
Acrab β Sco
Acrux α Cru
Acubens α Cnc
Adhafera ζ Leo
Adhara ε CMa
Adhil ξ And
Adib α Dra
Agena β Cen
Ain ε Tau
Albali ε Aqr
Albireo β Cyg
Alchiba α Crv
Alcor 80 UMa
Alcyone η Tau
Aldebaran α Tau
Alderamin α Cep
Alfirk β Cep
Algedi α Cap
Algenib γ Peg; α Per
Algieba γ Leo
Algol β Per
Algorab δ Crv
Alhena γ Gem
Alioth ε UMa
Alkaid η UMa
Alkalurops μ Boo
Alkes α Crt
Almach γ And
Almuredin ε Vir
Alnair ζ Cen
Al Na'ir α Gru
Alnasl γ Sgr

Alnilam ε Ori
Alnitak ζ Ori
Al Niyat σ Sco
Alphard α Hya
Alphecca α CrB
Alpheratz α And
Alrakis μ Dra
Alrami α Sgr
Alrescha α Psc
Alrucaba α UMi
Alshain β Aql
Altair α Aql
Altais δ Dra
Alterf λ Leo
Aludra η CMa
Alula Australis ξ UMa
Alula Borealis ν UMa
Alwaid β Dra
Alya θ Ser
Ancha θ Aqr
Ankaa α Phe
Antares α Sco
Arcturus α Boo
Arided α Cyg
Arkab β Sgr
Arneb α Lep
Ascella ζ Sgr
Asellus Australis δ Cnc
Asellus Borealis γ Cnc
Aspidiske ι Car
Atik o Per
Atlas 27 Tau
Atria α TrA
Avior ε Car

Azelfafage π^1 Cyg
Azha η Eri
Azimech α Vir

Baten Kaitos ζ Cet
Beid o^1 Eri
Bellatrix γ Ori
Benetnasch η UMa
Betelgeuse α Ori
Biham θ Peg
Botein δ Ari
Bungula α Cen

Canopus α Car
Capella α Aur
Caph β Cas
Castor α Gem
Cebalrai β Oph
Celaeno 16 Tau
Chara β CVn
Chertan θ Leo
Cor Caroli α CVn
Coxa θ Leo
Cujam ω Her
Cursa β Eri

Dabih β Cap
Dalim θ Eri
Deneb α Cyg
Deneb Algedi δ Cap
Deneb Kaitos β Cet
Denebola β Leo
Diphda β Cet
Dschubba δ Sco
Dubhe α UMa
Duhr δ Leo

Edasich ι Dra
Electra 17 Tau
Elnath β Tau
Eltanin γ Dra
Enif ε Peg
Errai γ Cep

Fomalhaut α PsA
Furud ζ CMa

Gacrux γ Cru
Garnet Star μ Cep
Gemma α CrB
Giausar λ Dra
Giedi α Cap
Gienah γ Crv; ε Cyg
Girtab θ Sco
Gomeisa β CMi
Graffias β Sco
Grumium ξ Dra

Hadar β Cen
Hamal α Ari
Heka λ Ori
Homam ζ Peg

Izar ε Boo

Kaus Australis ε Sgr
Kaus Borealis λ Sgr
Kaus Media δ Sgr
Keid o^2 Eri
Kitalpha α Equ
Kochab β UMi
Kornephoros β Her
Kurhah ξ Cep

Lesath υ Sco

Maasym λ Her
Maia 20 Tau
Marfik λ Oph
Markab α Peg
Markeb κ Vel
Marsic κ Her
Matar η Peg
Mebsuta ε Gem
Megrez δ UMa
Meissa λ Ori
Mekbuda ζ Gem

Menkalinan β Aur
Menkar α Cet
Menkent θ Cen
Menkib ξ Per
Merak β UMa
Merga h Boo
Merope 23 Tau
Mesarthim γ Ari
Miaplacidus β Car
Mimosa β Cru
Mintaka δ Ori
Mira o Cet
Mirach β And
Mirfak α Per
Mirzam β CMa
Mizar ζ UMa
Mothallah α Tri
Muhlifain γ Cen
Muliphein γ CMa
Muphrid η Boo
Muscida o UMa

Nair al Zaurak α Phe
Naos ζ Pup
Nashira γ Cap
Nekkar β Boo
Nihal β Lep
Nunki σ Sgr
Nusakan β CrB
Nushaba γ Sgr

Peacock α Pav
Phact α Col
Phecda γ UMa
Pherkad γ UMi
Pleione 28 Tau
Polaris α UMi
Pollux β Gem
Porrima γ Vir
Procyon α CMi
Propus η Gem
Proxima α Cen C
Pulcherrima ε Boo

Rasalas μ Leo
Rasalgethi α Her
Rasalhague α Oph
Rastaban β Dra
Regor γ Vel
Regulus α Leo
Rigel β Ori
Rigil Kentaurus α Cen
Rotanev β Del
Ruchbah δ Cas
Rukbat α Sgr
Rutilicus β Her

Sabik η Oph
Sadachbia γ Aqr
Sadalbari μ Peg
Sadalmelik α Aqr
Sadalsuud β Aqr
Sadr γ Cyg
Saiph κ Ori
Sargas θ Sco
Scheat β Peg
Seginus γ Boo
Sertan α Cnc
Sham α Sge
Shaula λ Sco
Shedar α Cas
Sheliak β Lyr
Sheratan β Ari
Sirius α CMa
Sirrah α And
Situla κ Aqr
Skat δ Aqr
Spica α Vir
Sterope 21 Tau
Sualocin α Del
Subra o Leo
Suhail λ Vel
Suhel α Car
Sulafat γ Lyr
Syrma ι Vir

Talitha   ι UMa
Tania Australis   μ UMa
Tania Borealis   λ UMa
Tarazed   γ Aql
Taygeta   19 Tau
Tegmine   ζ Cnc
Tejat   μ Gem
Theemin   43 Eri
Thuban   α Dra
Toliman   α Cen
Tureis   ϱ Pup

Unukalhai   α Ser

Vega   α Lyr
Vindemiatrix   ε Vir

Wasat   δ Gem
Wazn   β Col
Wezen   δ CMa

Yed Posterior   ε Oph
Yed Prior   δ Oph
Yildun   δ UMi

Zaniah   η Vir
Zaurak   γ Eri
Zavijava   β Vir
Zibal   ζ Eri
Zosma   δ Leo
Zubenelgenubi   α Lib
Zubeneschamali   β Lib

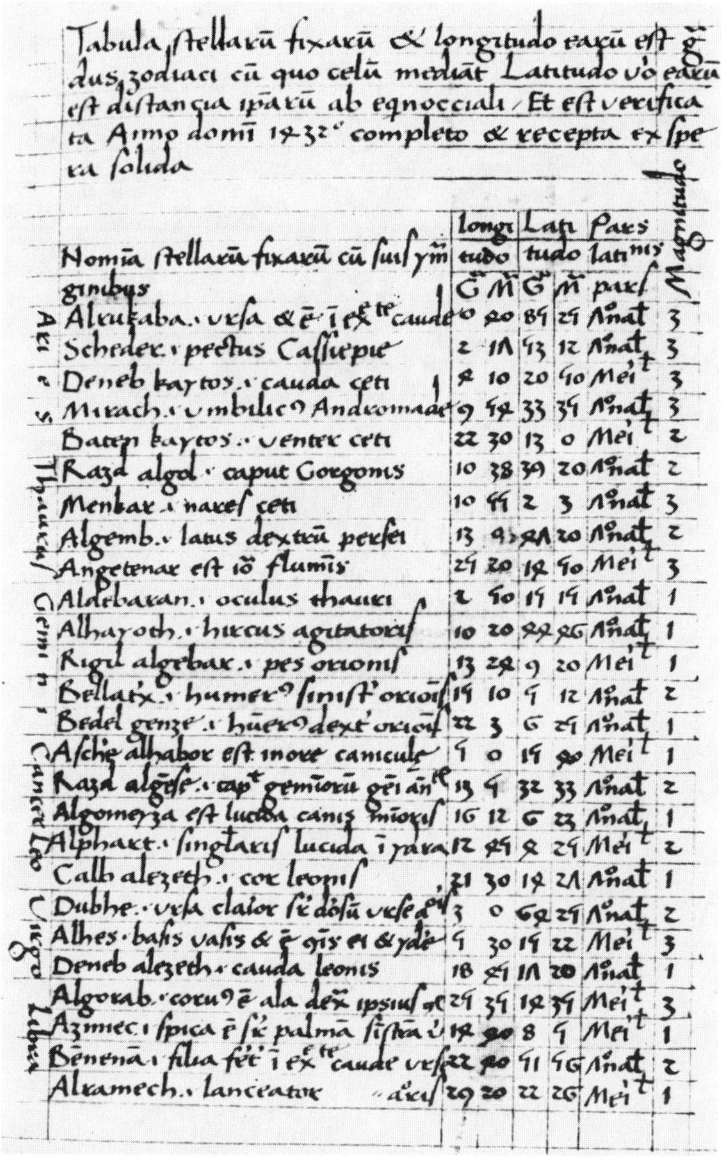

Fig. 3: Part of a star table established, in Vienna, for the end of the year 1432 (Nuremberg, Stadtbibliothek, MS Cent. VI 18). Arabic star names with Latin explanations. Many of these names are still used today. The table is in the hand of Regiomontanus. (Courtesy of Stadtbibliothek, Nuremberg)

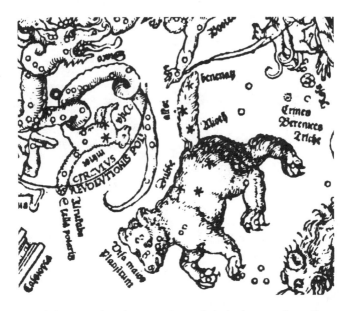

Fig. 4: Detail from a celestial map in Peter Apian's *Astronomicum Caesareum*, Ingolstadt, 1540, showing the classical 48 constellations and some star names. For Ursa Major there are marked *Dubhe*, *Alioth*, *benenatz* (afterwards spelled Benetnasch) and *alkor* (for 80 UMa; this name, spelled as *Alcor*, appears for the first time in two other books of Apian, in 1524). The Pole Star, in Ursa Minor, is called *Alrukaba* and *Stella polaris*. The region of Coma Berenices carries the designations *Crines Berenices* (contemporary), *Triche* and *Rosa* (both taken from the medieval Latin translation of the *Almagest* from Arabic). Two dogs, unnamed, are following Bootes; later they became a constellation of their own, Canes Venatici.